D0617258
.... AND HE WILL BE YET WISER Proverbs 9.9
ARABIC ASTRONOMY
CALCULUS
COMMERCIAL CORRESPONDENCE
CRICKET DRAWING
ELECTRICITY IN THE HOUSE
ENGLISH RENASCENCE TO THE
EVERYDAY FRENCH TO EXPRESS
BOOK GARDENING GAS IN THE HOUSE GEOGRAPHY OF
GERMAN GRAMMAR GERMAN PHRASE BOOK GOLF
GOOD FARM ACCOUNTING GOOD FARM CROPS GOOD FARMING
FARMING GOOD GRASSLAND GOOD AND HEALTHY ANIMALS
GOOD POULTRY KEEPING GOOD SHEEP FARMING GOOD SOIL
HINDUSTANI HISTORY: ABRAHAM LINCOLN ALEXANDER THE
CONSTANTINE COOK CRANMER ERASMUS GLADSTONE AND
MILTON PERICLES PETER THE GREAT PUSHKIN RALEIGH RICHELIEU
AND PURPOSE SOCCER SPANISH
SWEDISH TEACHING THINKING
RAILWAYS FOR BOYS CAMPING FOR BOYS AND GIRLS
FOR GIRLS MODELMAKING FOR BOYS NEEDLEWORK FOR GIRLS
BOYS AND GIRLS SAILING AND SMALL BOATS FOR BOYS AND GIRLS
FOR BOYS ADVERTISING & PUBLICITY ALGEBRA AMATEUR
BIOLOGY BOOK-KEEPING BRICKWORK BRINGING UP
CHEMISTRY CHESS CHINESE COMMERCIAL ARITHMETIC
TRAVELLING TO COMPOSE MUSIC CONSTRUCTIONAL DETAILS
DUTCH DUTTON SPEEDWORDS ECONOMIC GEOGRAPHY
EMBROIDERY ENGLISH GRAMMAR LITERARY APPRECIATION
ROMANTIC REVIVAL VICTORIAN AGE CONTEMPORARY
FISHING TO FLY FREELANCE WRITING FRENCH FRENCH
GEOGRAPHY OF LIVING THINGS GEOLOGY GEOMETRY
BOOK GOLF GOOD CONTROL OF INSECT PESTS GOOD
FARM CROPS GOOD FARMING GOOD FARMING BY MACHINE
GOOD AND HEALTHY ANIMALS GOOD MARKET GARDENING
GOOD SHEEP FARMING GOOD SOIL GOOD ENGLISH GREEK
ABRAHAM LINCOLN ALEXANDER THE GREAT BOLIVAR BOTHA
CRANMER ERASMUS GLADSTONE AND LIBERALISM HENRY V JOAN OF
PUSHKIN RALEIGH RICHELIEU ROBESPIERRE THOMAS JEFFERSON
HOME NURSING HORSE MANAGEMENT HOUSEHOLD DOCTOR
LATIN LAWN TENNIS LETTER WRITER MALAY
WORKSHOP PRACTICE MECHANICS MECHANICAL
MORE GERMAN MOTHERCRAFT MOTORING MOTOR CYCLING
PHYSICAL GEOGRAPHY PHYSICS PHYSIOLOGY PITMAN'S
PSYCHOLOGY PUBLIC ADMINISTRATION PUBLIC SPEAKING

THE TEACH YOURSELF BOOKS
EDITED BY LEONARD CUTTS

GEOMETRY

[E.N.A.

Geometrical forms and symmetry in Architecture.
The beautiful West Front of Florence Cathedral.

TEACH YOURSELF

GEOMETRY

By

P. ABBOTT, B.A.

Published in the U.S.A.
by
DOVER PUBLICATIONS, INC.
180 Varick Street
New York 14, New York

THE ENGLISH UNIVERSITIES PRESS LTD
102 NEWGATE STREET
LONDON, E.C.1

First printed March 1948
This impression 1959

Printed in Great Britain for the English Universities Press, Limited, by Richard Clay and Company, Ltd., Bungay, Suffolk

PREFACE

The primary object of this book is to provide an introduction to the fundamental principles of Geometry suitable for a private student, whether he be one who is desirous of beginning the study of the subject or one who, after a compulsory gap in his education, wishes to refresh his memory of previous studies.

The general plan of the book, modified in accordance with its special purpose, follows, in the main, recommendations made some years ago by the Teaching Committee of the Mathematical Association, of which committee the writer was at the time the Hon. Secretary. Accordingly there is a first part which is intended to lead the student to a realization of basic geometric truths by appealing to common sense reasoning and intuition. The usual proofs, when introduced are considerably modified, the formal proofs in logical sequence being postponed to Part II. The use of geometry in our everyday life is constantly indicated so that the student does not feel that the subject is merely one of academic interest.

Very little " practical geometry," involving drawing and measurements, is employed, as it is thought to be hardly suitable to the kind of student for whom the book is written. When, however, the theorems enunciated are suitable for the purpose, a considerable number of numerical exercises are included, their main purpose being to impress the theorems on the memory. Also such elementary mensuration as arises naturally from the geometry is introduced and the student thus acquires a knowledge of the ordinary rules for the calculation of areas and volumes.

No previous knowledge of Mathematics, beyond ordinary Arithmetic, is required by a student who proposes to use the book. It is desirable, however, from every point of view that the student who possesses but little knowledge of algebra should begin his study of that subject concurrently. At a later stage, Trigonometry should be started when the student will begin to find himself weaving together threads from all three subjects and realising their interdependence.

CONTENTS

CHAPTER 5

DIRECTION

CHAPTER 6

TRIANGLES

CHAPTER 7

PARALLEL STRAIGHT LINES

CHAPTER 8

ANGLES OF A TRIANGLE

CHAPTER 9

ISOSCELES TRIANGLES

CHAPTER 10

FUNDAMENTAL CONSTRUCTIONS

CHAPTER 11

QUADRILATERALS. PARALLELOGRAMS

CHAPTER 12

AREAS OF RECTILINEAL FIGURES

CHAPTER 13

THEOREM OF PYTHAGORAS

CHAPTER 14

POLYGONS

CHAPTER 15

LOCI

CHAPTER 16

THE CIRCLE

CHAPTER 17

THE CIRCLE (*contd.*)

CHAPTER 18

THE CIRCLE (*contd.*)

CHAPTER 19

CIRCLE. TANGENTS

CHAPTER 20

RATIO IN GEOMETRY. SIMILAR FIGURES

CHAPTER 21

EXTENSION OF THE THEOREM OF PYTHAGORAS

CHAPTER 22

SYMMETRY

CHAPTER 23

SOLID GEOMETRY

CHAPTER 24

PRISMS

CHAPTER 25

PYRAMIDS

CHAPTER 26

SOLIDS OF REVOLUTION

PART II

ABBREVIATIONS

The following abbreviations are used occasionally throughout this book.

Sign.	Meaning.
$=$	is equal to.
$>$	is greater than.
$<$	is less than.
$\parallel$	is parallel to.
$\angle$	angle.
$\triangle$	triangle.
sq.	square.
$\parallel$ gram	parallelogram.
rect.	rectangle.
rt.	right.
$\therefore$	therefore.

INTRODUCTION

WHAT IS GEOMETRY?

1. The Practical Origin of Geometry.

The word " geometry " is derived from two Greek words, and means "*earth measurement.*" This suggests that in its beginnings the subject had a practical basis, with which the Greeks were familiar. It is known that the Greeks did not originate geometry, but became acquainted with the subject through their intercourse with the Egyptians, who, by tradition, were the first to develop the science. Ancient inscriptions and records indicate that this gifted race employed some of the principles of geometry in land surveying, together with simple developments, such as are now included in the subject of Trigonometry.

This practical application of geometry appears to have originated in the annual recurrence of widespread floods in the Nile valley. These resulted in the obliteration of many of the boundaries of private lands. Hence the necessity of restoring them after the subsidence of the waters of the river. Originally the work was undertaken by the priests; to accomplish it they applied certain geometrical principles, many of which they no doubt discovered.

It is also a fair assumption that the construction of their massive temples, tombs and pyramids could scarcely have been accomplished without a considerable knowledge of geometry and mechanical principles.

2. The Development of Abstract Geometry by the Greeks.

It was, however, the abstract conceptions and logical reasoning of geometry which made a special appeal to the Greeks: to them, abstract reasoning of any kind was congenial. Consequently, when philosophers adopted geometry as a subject for study and discussion, they were not satisfied with the knowledge of some geometrical truth; they sought for logical and incontrovertible proof of it.

Gradually there came into being a considerable body of geometric theorems, the proofs of which were known and were parts of a chain of logical reasoning. The proof of any particular theorem was found to be dependent on some other theorem or theorems, and logically could not be based on them unless the truth of these, in their turn, had been established. Nothing was to be assumed, or taken for granted, except certain fundamental self-evident truths, termed *axioms,* which from their nature were usually incapable of proof.

Thus there gradually was established a body of geometrical knowledge forming a chain of geometrical reasoning in a logical sequence.

3. Euclid's Sequence.

As far as is known, one of the first mathematicians to formulate such a logical sequence was Euclid, who was born about 330 B.C. His book, which incorporated work accomplished by previous writers, was one of the most famous ever written upon a mathematical subject. It became the recognised text-book on elementary geometry for some two thousand years, extending to our own times. In modern times it has been displaced by a variety of text-books, intended to render the subject more in accordance with the needs of the age. These books, in varying degrees, avoid what are now recognised as defects in Euclid, introduce changes in the sequence and incorporate such new topics and matter as modern opinion and necessity demand. It has also become usual that a course in practical geometry should precede the study of abstract formal geometry. This method, with some modifications, has been followed in this book.

4. The Practical Aspects of Geometry.

We have seen that in its origins geometry was essentially a **practical** subject. This aspect of it has of necessity continued to be of increasing importance throughout the centuries, since it is essential in all draughtsmanship necessary in the work of engineers, architects, surveyors, and others.

Practical geometry, in this sense, is mainly concerned with the construction of what may be termed geometrical figures. Some of the simpler of these constructions have always been included in the abstract logical treatment of the subject, the accuracy of the methods employed being proved theoretically. For example, in practical geometry we may learn the mechanical method of bisecting a straight line, and go no farther. But the method is made evident by theoretical geometry, and has been proved logically and conclusively to produce the desired result.

A knowledge of the fundamental principles of geometry is also necessary for the study of other branches of mathematics, such as trigonometry and mechanics—subjects which are of vital importance to engineers of all kinds as well as to those who are proceeding to more advanced work in mathematics.

5. The Treatment of Geometry in this Book.

Geometry may thus be treated from two aspects:

(1) The practical applications of the subject, and

(2) As a method of training in mathematical and logical reasoning.

These are reflected in the plan of this book, which consists of two parts:

Part I. This will be concerned with the investigation and study of the salient facts of elementary geometry, practical methods, intuition and deduction being freely employed to demonstrate their truth. It is designed to enable the student more easily, and with more understanding, to proceed to a full and logical treatment of the subject.

Part II consists of a short course of formal abstract geometry. Limitations of space do not allow of a full treatment, but it is hoped that it will be sufficient to enable the student to realise the meaning, and perhaps to feel something of the satisfaction to be derived from the logical completeness of mathematical reasoning, from which vague, unsupported statements and loose thinking are excluded.

PART I

PRACTICAL AND THEORETICAL GEOMETRY

CHAPTER I

SOLIDS, LINES AND POINTS

1. Geometric Forms and Figures.

It is seldom realised to what an extent the terms and facts of geometry are woven into the fabric of our daily life, and how important is the part they play in our environment. Such geometric terms as square, rectangle, straight line, circle and triangle are familiar to everybody. Most people realise what they signify, though ideas about them may occasionally be vague and lacking in precision.

We are familiar also with the pictorial representation of

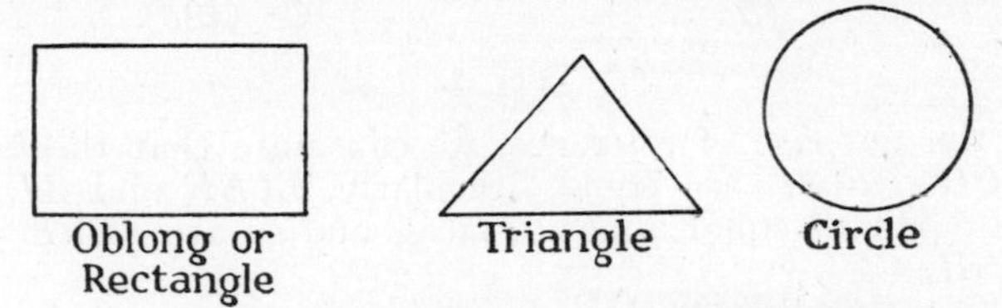

FIG. 1.

these terms by means of drawings such as are shown in Fig. 1. These drawings we may call **geometric figures.** They will be found very useful when examining and discussing the properties of the particular forms represented.

2. Geometric Figures and Solids.

Many of the geometric figures which we see around us are surfaces of what are termed **solid bodies.** As an example, examine the outside cover of an ordinary box of matches.

This box has **six sides or faces**, each of which is an **oblong** or a **rectangle**.

This box may be represented by a drawing in two ways. In Fig. 2 (*a*), it is drawn as we see it. Owing to the wood of which the box is composed, **three** faces only of the box are visible. But for the examination of the figure from the point of view of geometry, it is usually drawn as shown in Fig. 2 (*b*). There it is represented as though the solid were a kind of skeleton, constructed with fine wires, so that all the faces can be seen. Those which are actually hidden, in reality, are represented by dotted lines.

In this form we are better able to examine the construction of the body and to develop relations which exist between parts of it. For example, attaching letters to the corners

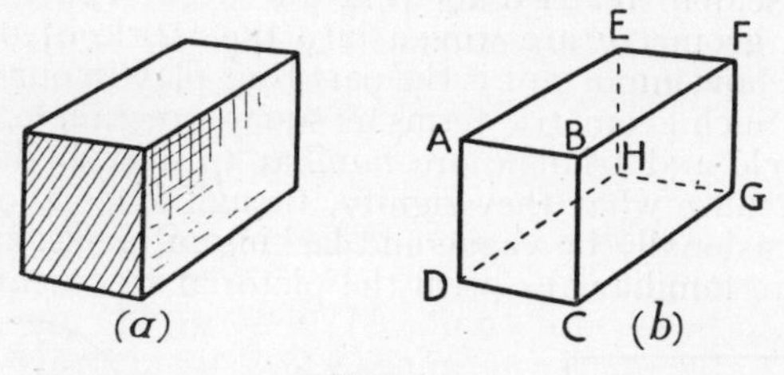

FIG. 2.

for the purpose of reference, we can state that the faces *ABCD*, *EFGH*, are equal. Similarly, *BCGF* and *ADHE* are a pair of equal opposite faces, and so are *ABFE* and *DCGH*.

We spoke above of this as a **solid body**, and we must dwell for a moment on the sense in which the term is used in geometry. In the ordinary way we mean by the term "solid" something which is compact with matter. But in geometry we are concerned only with a portion of space enclosed or bounded by surfaces, and are not concerned with the matter or material which it might or might not contain. We think only of the abstract shape of the solid. Thus :

A solid body from the point of view of geometry is conceived as occupying space, and the amount of this space is called its volume.

3. Surfaces, Lines, Points.

Examining in more detail the box represented in Fig. 2, we note the following points:

(1) The box is bounded or enclosed by **six faces or sides,** which we call **surfaces.**

(2) **Two adjacent faces meet in a straight line,** which is called an **edge.** Thus the faces $ABCD$ and $FBCG$ meet in the straight line BC. In the whole solid there are twelve of these edges.

(3) **The intersection of two edges is a point.** For example, the edges AB and BC meet in a point which is indicated by B. There are eight such points, commonly referred to as corners. Each of these also indicates the meeting point of three edges. Thus B marks the intersection of the edge BF with the edges AB and BC.

4. Definitions.

In the preceding section three geometric terms occur: **surface, straight line, point.** It is very important, when geometric terms are employed, that we should be quite clear as to the precise meanings which are attached to them. It is necessary, therefore, that such terms should be clearly and accurately defined.

Before proceeding to deal with definitions of the terms above, it is desirable that we should consider for a moment what should constitute a clear and accurate definition. At a later stage this will be dealt with more fully, but it may be stated now that *definitions should employ no words which themselves require definition.* Further, *they should contain no more words or statements than are necessary for accurate description.*

There are terms in geometry, however, which describe fundamental notions, for which no satisfactory definitions have been framed, or are possible. They are terms for which no simpler words can be found, and at the same time are so clearly understood by everybody that definitions are not really necessary; there is no misconception as to their meaning. Among such terms are those employed above—viz., **points, straight lines** and **surfaces.**

In the same category as these, are many other words outside geometry in everyday use, such as colour, sweet, noise and shape, which we cannot define by the use of simpler words, but we know exactly what they mean.

In geometry, though we may not be able to define certain terms, such as those employed above, it is necessary to examine further the sense in which they are employed, when they occur in the subject.

5. Points.

It was stated in § 3 that the **edges AB and BC of the box meet in the point B.** This means that **the point B marks the position in space where the straight lines AB and BC meet.** It is a familiar act with all of us to mark a position on a piece of paper, or a map, or on a picture by making a small dot, and we speak of that as showing some particular position which we wish to indicate. Thus we may say that

A point indicates position in space.

Although we make a small dot, which is visible, to mark a particular position, in theory **a point has no size or magnitude.** Sometimes, for various reasons, we make a small

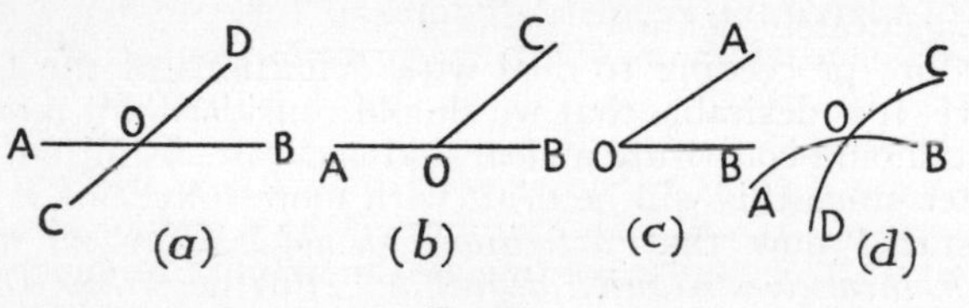

Fig. 3.

cross instead of a dot to indicate position, and in that case **the point lies at the intersection of the two lines** forming the cross.

In Fig. 3 is shown the position of a point as marked by:

(*a*) The intersection of two straight lines, *AB* and *CD*, at *O*.

(*b*) The **meeting** of two straight lines, *AB* and *OC*, at *O*.

(*c*) The meeting of two straight lines, *OA* and *OB*, at *O*.

(*d*) The intersection of two curved lines, *AB* and *CD*.

The student should note the differentiation in the above between a straight line and a curved line.

6. A Straight Line.

It was stated in § 3 that when two faces of the solid intersected a **straight line** was formed. We were thus using the term " straight line " before defining it. No confusion or misunderstanding is caused thereby, because everybody knows what is meant by a straight line, though no satisfactory definition of it has been formulated. However, it is necessary to investigate further the term as it is used in geometry.

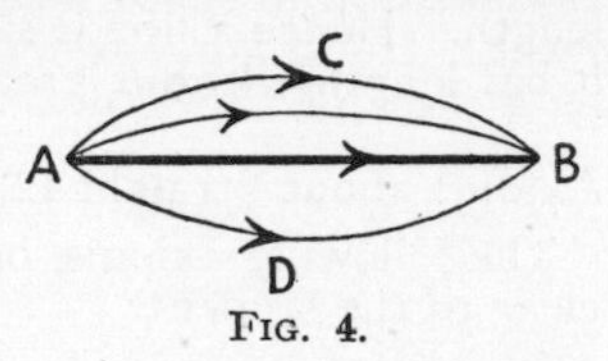

FIG. 4.

Straight lines occur in very many other ways besides the intersection of two faces of a solid. They were employed, for example, in the construction of two of the geometric figures of Fig. 1. They enter into the constructions of the majority of geometric figures.

There is a further way in which the formation of a line may be imagined.

Suppose a point to move along the surface of the paper, or in space. It will mark out a line, which may be straight, curved or irregular, according to the manner in which it is moving.

In Fig. 4 let *A* and *B* be two points on the surface of the paper. Imagine a point at *A* to move to the position *B*. There is an innumerable number of paths which it may take, such as those indicated by *ACB* and *ADB*. These vary in length, but we know intuitively that the most direct way will be along the **straight** line *AB*, which joins the points. Just as, if we wish to cross a field from one side to the other, the nearest and quickest way, other things being equal, is along a **straight** path.

Thus we arrive at a **description** of a straight line as "*the shortest distance between two points.*"

It will be noted that this idea of a line being formed as the path of a moving point is illustrated in drawing, when the **point of the pencil** moves along the paper, either along a ruler or straight edge to produce a straight line, or guided by the compass to form a circle.

It was stated in § 5 that a point has no size or magnitude. Consequently the *straight line which marks the path of a moving point can have no width,* though when drawing a representation of it on paper a slight width is given to it, in order to make it visible to the eye.

It can, however, be measured in one way only—*i.e.*, its length. Hence a line is said to be of one dimension only; it has length without breadth.

Axioms about Straight Lines and Points.

The following axioms, or self-evident truths, will now be clear to the student:

(1) *One straight line only can be drawn to pass through two points.*

(2) *Two straight lines can intersect in one point only.*

(3) *Two straight lines cannot enclose a space.* Otherwise, they must meet in more than one point.

7. Surface.

It was pointed out in § 3 that the box which we were considering—a solid—was marked off, or bounded, from the surrounding space by **six faces** or **surfaces.** This is true for all solids; the space which they occupy is bounded by surfaces.

As stated previously, "**surface**" is another geometric term which cannot be satisfactorily defined; but every student will understand the meaning of it. You write on the surface of a sheet of paper, you polish the surface of a table; you may observe the surface of the water in a tumbler.

Area of a Surface. If it is required to find the size or magnitude of a given surface, as for example the page you are reading, we must know both the **length** and **breadth** of it, since the size of it evidently depends on both of these. This will be found to be true of the surfaces covered by all regular geometric figures; **two measurements are necessary.** Hence a **surface is of two dimensions.** Evidently thickness or depth does not enter into the conception of a surface. **The amount of surface covered by a figure such as a rectangle or circle is called its area.**

8. Plane Surfaces.

Some surfaces are perfectly flat or level, such as the surface of the paper on which this is printed, or the top of a polished table, or the surface of still water.

Such surfaces are called plane surfaces, or, more briefly, **planes.** No formal definition of a plane surface can be given, but the meaning of a flat or level surface is perfectly clear to everybody.

Other surfaces may be curved, such as that of the sides of a jam jar, a billiard ball, etc., but for the present we are not concerned with these.

Test of a Plane Surface. A plane surface could be tested as follows:

If any two points are taken on the surface, the straight line which joins them lies wholly in the surface.

Acting on this principle, a carpenter tests a surface, such as that of a piece of wood, which he is " planeing " to produce a level surface for a table, etc.

This is clearly not the case with curved surfaces. If, for example, you take a rubber ball and make two dots on its surface some distance apart, it is obvious that the straight line joining them would not lie on the surface of the ball. They can be joined by a curved line on the surface of the ball, but that will be discussed in a later chapter.

We shall return to plane surface or planes later, but for the present we shall proceed to discuss figures which lie in a plane. Such figures are called **plane figures.**

CHAPTER 2

ANGLES

9. When two straight lines meet they are said to form an angle.

Or we may say that they **include** an angle.

This is a **statement** of the manner in which an angle is formed; it is not a definition; and, indeed, no satisfactory definition is possible. It is incorrect to say that the angle is the space between two intersecting lines: we have seen that two straight lines cannot enclose a space.

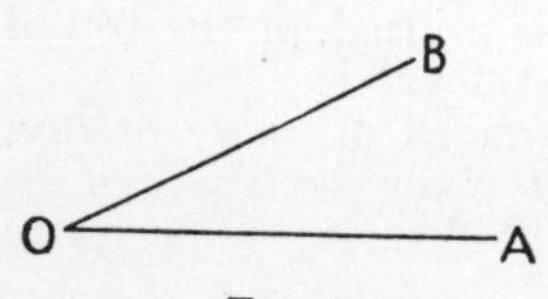

FIG. 5.

Arms of an Angle. The straight lines which meet to form an angle are called the "arms" of the angle.

Vertex. The point where the two arms meet is called the vertex.

In Fig. 5 *AOB* represents the angle formed by the meeting of the two arms *OA* and *OB*, and *O* is the vertex.

It is evident that the *size* of the angle does not depend on the lengths of the arms; this will be seen if the arms *OA*, *OB* in Fig. 5 are produced. In making a drawing to scale, whether the drawing is reduced or enlarged, **all angles remain the same in size.**

Naming an Angle. When letters are employed to denote an angle, it is usual to use three, as the angle *AOB* in Fig. 5, the middle letter being that which is placed at the vertex. Then *OA* and *OB* represent the arms of the angle. When there can be no doubt as to the angle referred to, the letter at the vertex—*O* in Fig. 5—is often used by itself to denote the angle; thus, we may speak of the angle *O*.

The phrase "the angle *AOB*" may be abbreviated to $\angle AOB$ or $A\hat{O}B$.

10. Adjacent Angles.

When a straight line meets two other straight lines, as *AO* meets *CO* and *DO* in Fig. 6, two angles are formed, with a common vertex *O*. These angles, *AOD*, *AOC*, are called **adjacent angles.** If *CO* be produced to *B*, then ∠s *COA*, *BOA* are adjacent angles. So also are ∠s *AOD*, *BOD* and the ∠s *COD*, *BOD*.

Definition. *Angles which have a common vertex, one common arm and are on opposite sides of the common arm, are called adjacent angles.*

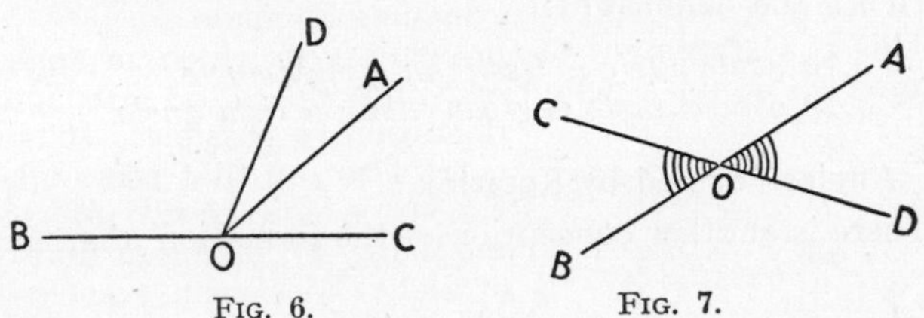

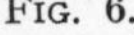
Fig. 6.

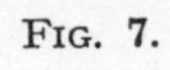
Fig. 7.

When two lines intersect, as in Fig. 7, four pairs of adjacent angles are formed.

11. Vertically Opposite Angles.

When two straight lines cut one another, as *AB* and *CD*, which intersect at *O*, in Fig. 7, the two angles *AOD*, *BOC* are called **vertically opposite angles.** The other two angles which are formed—viz., *AOC*, *BOD*—are also vertically opposite angles. Such angles have a common vertex.

Fig. 8.

12. Right Angles.

When a straight line such as *AO* meets another straight line *BOC* (Fig. 8) then, *if the adjacent angles are equal, each of these is called a right angle.*

AO is then said to be **perpendicular** to *BC*, and *BC* is perpendicular to *AO*.

13. Acute and Obtuse Angles.

In Fig. 8 the straight line *OD* is drawn to meet the straight line *BOC* at *O*, thus forming the angles *DOC* and *DOB* with *BC*. It is evident that of these two angles:

$\angle DOC$ is **less** than a right angle. It is called an **acute** angle.

$\angle DOB$ is **greater** than a right angle, and is called an **obtuse** angle.

Hence the **definitions:**

An acute angle is less than a right angle.
An obtuse angle is greater than a right angle.

14. Angles Formed by Rotation.

There is another conception of the formation of an angle

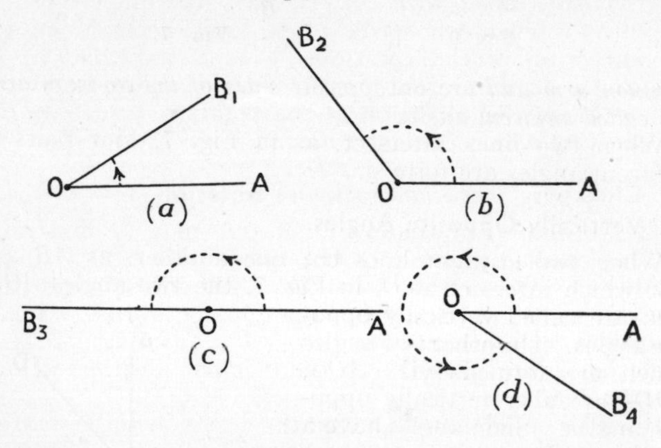

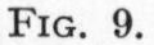
Fig. 9.

which is of great importance in practical applications of mathematics.

Take a pair of compasses, and keeping one arm fixed, as *OA* in Fig. 9, rotate the other arm, *OB*, slowly. As the moving arm rotates it forms with the fixed arm a succession of angles which increase in magnitude. In Fig. 9 are shown four of these angles, AOB_1, AOB_2, AOB_3 and AOB_4. Considering these angles, it is noted that:

In (*a*) the angle AOB_1 is an **acute** angle;

In (*b*) the angle AOB_2 is an **obtuse** angle, and

In (*c*) the rotating arm is in the **same straight line** with the fixed arm *OA*. Although this seems to be inconsistent with the idea of an angle in § 9, nevertheless it is formed in the same way as the acute and obtuse angles, and so AOB_3 **must be regarded as an angle, formed by rotation.** This is sometimes called a **straight angle,** and it will be considered again later.

(*d*) Continuing the rotation beyond the straight angle a position such as AOB_4 is reached. Such an angle, greater than a straight angle, is called a **reflex angle.** It must not be confused with the acute angle which is also formed with *OA*. Clearly the angle which is meant when we speak of $\angle AOB_4$ depends on the **direction of the rotation.** This is indicated by an arrow on the dotted curve. It is therefore important to know **the direction of the rotation** before we can be sure which angle is referred to.

15. Clockwise and Anti-clockwise Rotation.

The formation of angles by rotation may be illustrated by the familiar example of the hands of a clock. If the rotation of the minute hand be observed, starting from twelve o'clock, all the above angles, acute, obtuse, straight and reflex, will be formed in turn. For example, a straight angle has been formed with the original position at half-past twelve.

It will be noted, however, that the direction of the rotation is opposite to that indicated in Fig. 9. This movement is from left to right, whereas the minute hand moves right to left.

When the direction of the rotation is the same as that of the hands of a clock it is called **clockwise,** but when in the opposite direction, **anti-clockwise.** Thus if the angle AOB_4 (Fig. 9 (*d*)) is formed by clockwise rotation, it is an **acute** angle, if by anti-clockwise, **reflex.**

Mathematically, **anti-clockwise rotation** is conventionally regarded as a standard direction and considered to be **positive,** while clockwise rotation is considered as **negative.**

16. Rotating Straight Lines.

We must now proceed to examine the idea of rotation in the abstract by imagining the rotation of a straight line.

Suppose a straight line *OA* (Fig. 10) to start from a fixed position to rotate in the plane of the paper about a fixed point *O* on the line, the rotation being in an anti-clockwise direction.

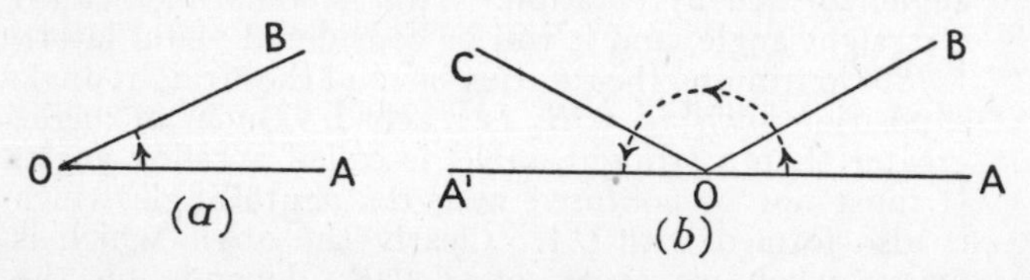

FIG. 10.

When it has reached any position such as *OB* (Fig. 10 (*a*)), an angle *AOB* has been formed by it with the original position *OA*.

Thus we have the conception of *an angle as being formed by the rotation of a straight line about a fixed point on it, which becomes the vertex of the angle.*

As the rotation continues to another position such as *OC* (Fig. 10 (*b*)), an obtuse angle, *AOC*, is formed. If the rotation is continued, the position *OA'* is reached, in which *A*, *O*, *A'* are in the same straight line.

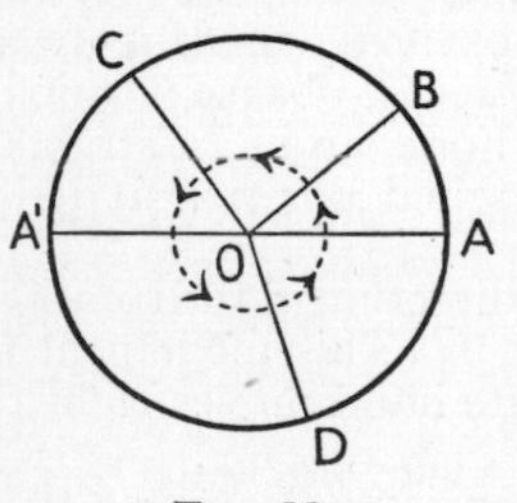

FIG. 11.

A Complete Rotation. Continuing the rotation, as shown in Fig. 11, the straight line passes through a position such as *OD*, and finally returns to *OA*, the position from which it started. **The straight line has thus made a complete**

rotation or revolution about the fixed point O which is the **centre of rotation.**

A Half Rotation. It is evident that when the position OA' is reached the rotating line has moved through **half a complete rotation.** OA and OA' are now in the same straight line. Hence the name straight angle (§ 14).

Reflex or Re-entrant Angle. When the rotating line reaches a position such as OB, shown in Fig. 12—that is, between a half and a complete rotation—the angle so formed is a **reflex** or **re-entrant** angle. The dotted curve and arrows indicate how the position has been reached (see § 14 (*d*)).

Angles of Unlimited Size. The student will probably

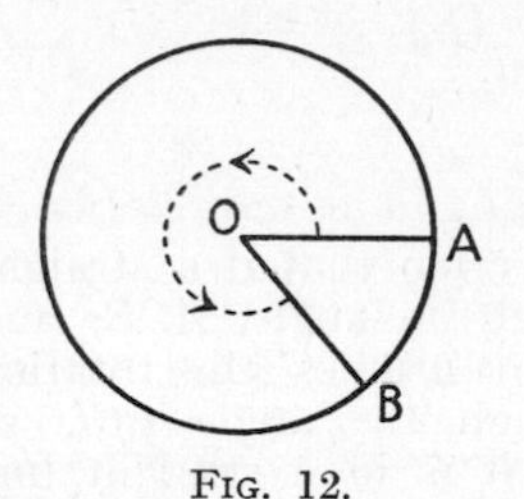

FIG. 12.

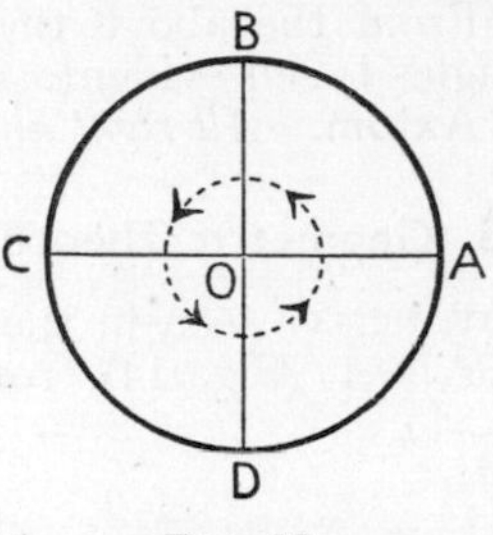

FIG. 13.

have noticed that the rotating line, after describing a complete rotation, may continue to rotate. In doing so it will pass again through all the positions indicated in Figs. 11 and 12 and go on to make two complete rotations. In this way the minute hand of a clock makes twenty-four complete rotations in twenty-four hours, while the hour hand makes two complete rotations in the same period. Clearly there is no limit to the possible number of rotations, and therefore, from this point of view, no limit to the size of an angle.

17. Right Angles and Rotation. The conception of an angle as being formed by rotation leads to a convenient method of describing a right angle.

Let the straight line OA (Fig. 13) describe a complete

rotation as indicated by the dotted curve. In the position *OC*, half a complete rotation has been made (§ 16).

Let *OB* be the position half-way between *OA* and *OC*. Then with equal amounts of rotation the two angles *AOB*, *BOC* will have been described.

Hence the angles AOB, BOC must be equal and are therefore right angles (§ 12).

Similarly, considering the position at *OD*, half-way in the rotation from *OC* onward to *OA*, the **angles COD and AOD must also be right angles**, and *BO* and *OD* must be in the same straight line.

Thus a complete rotation covers four right angles and a half rotation two right angles. Or the straight angle *AO*, *OC* equals two right angles.

From the above the following axiom relating to right angles is self-evident.

Axiom. *All right angles are equal.*

18. Geometric Theorems.

We have seen in the previous section that if a straight line, *OA* (Fig. 14), rotates through an angle, *AOB*, and then continues the rotation through the angle *BOC* so that it is in a straight line with its initial position *OA*, it has completed a half rotation. Consequently **the sum of the two angles must be two right angles** (§ 17). The angles *AOB*, *BOC* are adjacent angles (§ 10)—*i.e.*, they are the angles made when *OB* meets *AC*. The conclusion reached may be stated more concisely as follows:

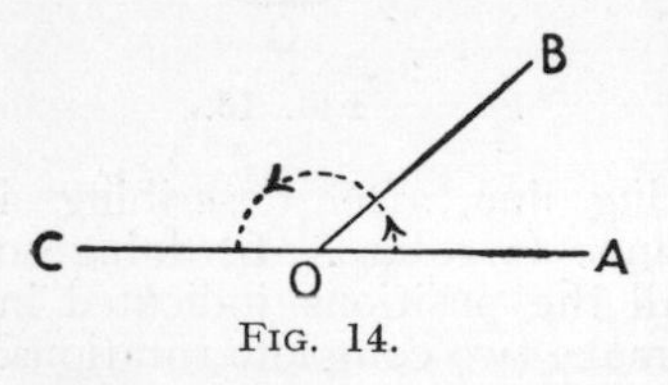

FIG. 14.

If the straight line OB meets the straight line AC at O, the sum of the angles so formed, AOB and BOC, is two right angles.

This is a statement of a geometric fact, and when expressed in general terms is called a geometric theorem. In such a form it would be stated thus :

Theorem. If one straight line meets another straight line, the sum of the two adjacent angles on one side of it is two right angles.

In this particular case the student, after reading the last few sections, will probably be satisfied as to the truth of the theorem but in general, a theorem cannot be accepted as being true until it has been proved to be so by methods of geometric reasoning.

The first step towards this is a clear and accurate statement of what has to be proved and what are the data from which we start. Thus in the above theorem the facts, which are given, are that one straight line (*OB* in Fig. 14) meets another straight line (*AOC*), and so forms two adjacent angles (*BOA*, *BOC*).

What has then to be proved is that the sum of these angles is two right angles.

There are thus two distinct parts of the theorem, and of all others.

(1) What is **given**—*i.e.*, the **data,** sometimes called the **hypothesis**—and

(2) **The proof.**

When the theorem has been stated in general form it is customary to draw a figure by means of which the two parts of the theorem can be clearly stated with special reference to this figure. By the use of this figure the proof of the theorem is developed.

19. Converse Theorems.

If the data and the proof are interchanged, we get a new theorem which is called the **converse** of the first theorem.

Applying this to the above theorem (1) it may be given that a straight line such as BO in Fig. 15 meets two other straight lines such as *CO* and *AO*, and that **the sum of the adjacent angles so formed—viz., BOC, BOA—is two right angles.** These are the **data** or **hypothesis.**

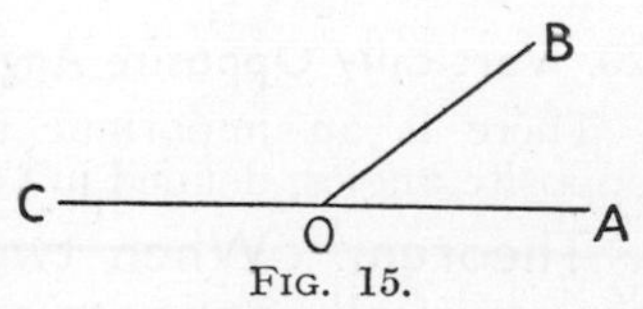

FIG. 15.

(2) **We then require to prove that CO and AO are in the same straight line**, or, in other words, *C*, *O* and *A* are in the same straight line.

This new theorem may be expressed in general terms as follows:

Theorem. If at a point in a straight line two other straight lines, on opposite sides of it, make the two adjacent angles together equal to two right angles, these two straight lines are in the same straight line.

The two theorems above are converse theorems. The hypothesis in the first theorem is what has to be proved in the second and vice versa.

It is important to remember that the converse of a theorem is not always true. Examples will occur in Part II.

It was stated above that a theorem cannot be accepted as being true until it has been proved to be so by geometrical reasoning. This will be adhered to in the formal treatment of the subject in Part II of the book, but in Part I, for various reasons, the strict proof will not always be given, especially with such theorems as those above, which will probably be accepted by the student as self-evident or axiomatic. They arise naturally from the conception of angles, and especially right angles, as being formed by the rotation of a straight line, as in § 16.

If the student desires to see how the above theorems can be proved he should turn to the proofs of Theorems 1 and 2 in Part II.

20. Vertically Opposite Angles.

There is an important theorem concerning vertically opposite angles, defined in § 11, which may be stated thus:

Theorem. When two straight lines intersect, the vertically opposite angles are equal.

The theorem is illustrated by Fig. 16, in which two straight lines *AB* and *CD* intersect at *O*, forming as shown, in § 11, two pairs of vertically opposite angles. It will be sufficient if this is proved to be true for one pair of angles only, say *COB*, *AOD*.

It is thus **required to prove** that $\angle COB = \angle AOD$.

Proof. In the Theorem of § 18 it was shown that:

(1) The adjacent $\angle$s, $\angle COB + \angle AOC = 2$ right $\angle$s.
and (2) „ „ $\angle AOD + \angle AOC = 2$ right $\angle$s.

But things equal to the same thing are equal to one another.

$$\therefore \quad \angle COB + \angle AOC = \angle AOD + \angle AOC.$$

Subtracting $\angle AOC$, which is common to both, the remainders must be equal—*i.e.*, $\angle COB = \angle AOD$.

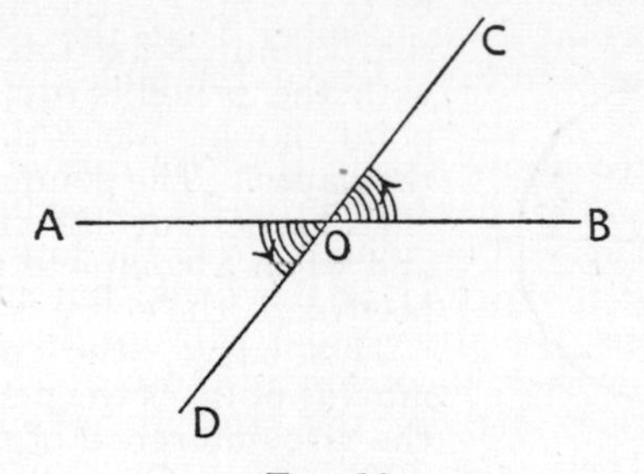

FIG. 16.

Similarly, the other pair of vertically opposite angles, *AOC* and *BOD*, may be proved equal.

The student is advised to write out this proof from memory as an exercise.

Proof by Rotation.

The equality of the angles is also evident by using the method of **rotation**.

Suppose the straight line *AOB* to rotate about *O* in an anti-clockwise direction to the position *CD*.

Then each arm must move through **the same amount of rotation.**

$$\therefore \quad \angle COB = \angle AOD.$$

CHAPTER 3

MEASUREMENT OF ANGLES

21. The Circle.

In Fig. 1, one of the geometric figures depicted is the familiar one known as a **circle.** It is a closed figure, bounded by one continuous curve. Mechanically it is constructed by using a pair of compasses, and the method, though well known, is recapitulated here. The arms having been opened out to a suitable distance represented by OA, Fig. 17, the arm with the sharp point is fixed at O and that with the pencil is rotated with its point moving along the surface of the paper. The point of the pencil then marks out the curve $ABCD$, and when a complete rotation has been made the curve is closed at A.

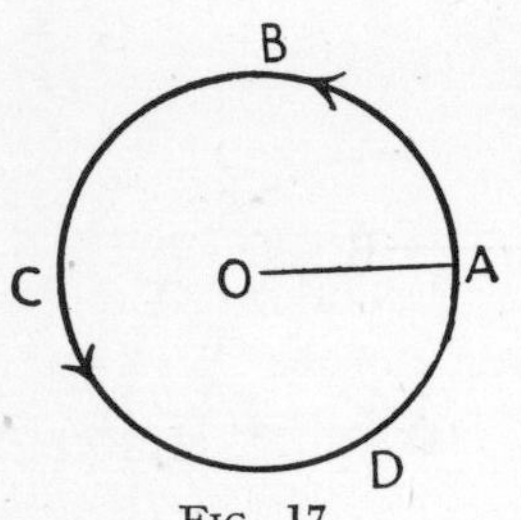

Fig. 17.

This curve, **the path of the moving point of the pencil, is called the circumference of the circle.**

The point **O is called the centre of the circle,** and the distances of all points on the circumference from O are equal. This distance, OA, is called the **radius** (plural radii).

A circle may now be defined as follows:

Definition. *A circle is a plane figure bounded by a curved line, called the circumference, and is such that all straight lines drawn from points on the circumference to a fixed point within the curve, called the centre, are equal.*

The circle may also be conceived as the **area** marked out in a plane by the rotation of a straight line, OA, about a point, O, at one end of the line.

Note.—The term " circle " is sometimes applied to the curved part—*i.e.*, the circumference—when there is no doubt what is meant, but, strictly, it is the name for the **whole figure.** Thus the **area of a circle,** as suggested above, is the area of that part of the plane which is enclosed by the circumference.

Arc of a Circle.

A part of the circumference is called an arc. Thus in Fig. 17 the part of the circumference between the points *B* and *C* is an arc.

(Other definitions connected with the circle are given in Chapter 16.)

Concentric Circles.

Circles which have the same centre but different radii are called concentric.

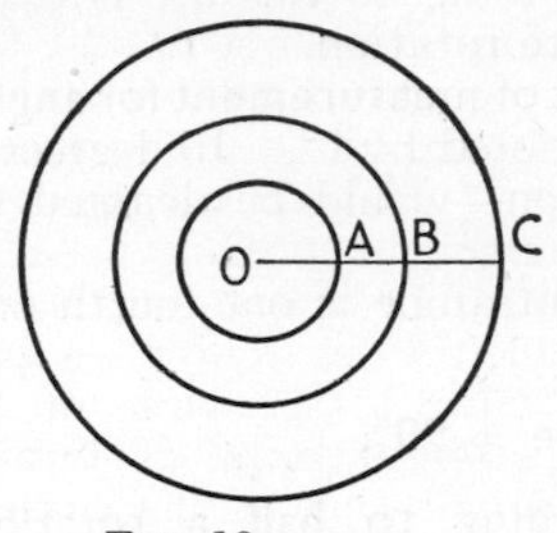

FIG. 18.

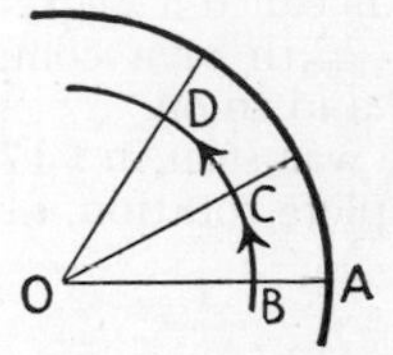

FIG. 19.

In Fig. 18, with centre *O*, and different radii, *OA*, *OB*, *OC*, three circles are described. These are concentric.

22. Measurement of Angles.

The conception of the formation of angles by the rotation of a straight line (§ 16) leads to a convenient method of measuring them.

When a straight line, *OA*, rotates about a point, *O*, Fig. 19, any point, *B*, on it will always be at the same distance from *O*, and consequently will describe a circle, concentric with that described by *OA*, as shown in Fig. 19. When an angle such as *BOC* is described, **the point B has marked out an arc of a circle, BC.**

The length of the arc clearly depends on the amount of rotation, as also does the size of the angle. The same amount of rotation as before will produce the angle *COD* equal to the angle *BOC*.

* Then **the arc CD must clearly be equal to the arc BC.** Thus, the angle *BOD* being twice the angle *BOC*, **the arc BD will be twice the arc BC,** and so for other multiples.

We may conclude, therefore, that **the length of the arc will depend on the size of the angle.** If the angle be doubled, the arc is doubled; if the angle be halved, the arc is halved.

Suppose the circumference of the circle, when a complete rotation has taken place, to be divided into 360 equal parts. Then each arc is $\frac{1}{360}$th of the whole circumference; consequently the angle corresponding to this arc is $\frac{1}{360}$th of that marked out in a complete rotation.

This angle is employed as a unit of measurement for angles and is called a **degree.** It is denoted by $1°$. 15 degrees—*i.e.*, $\frac{15}{360}$th of a complete rotation—would be denoted by $15°$, and so on.

It was seen in § 17 that a right angle is one-fourth of a complete rotation, *i.e.*, of $360°$.

$$\therefore \text{a right angle} = 90°.$$

A **straight angle, corresponding to half a rotation, contains 180°.**

Fig. 20 shows a circle, centre *O*, in which the circumference is divided into 360 equal parts. The arcs are comparatively very small, and so are the corresponding angles which, for each arc of one degree, are formed by joining the ends of the arc to *O*. Any particular angle made with *OA* can be constructed by joining the appropriate point to *O*.

For example, $\angle AOE$ is an angle of $45°$, and $\angle AOF$ is $120°$.

The $\angle AOC$, the straight angle, represents $180°$.

The straight line *BOD* is perpendicular to *AOC*, and thus the angles of $90°$ and $270°$ are formed.

* For a proof of this see Part II, Theorem 46.

For angles smaller than one degree the following subdivisions are used:

(1) Each degree is divided into 60 equal parts, called **minutes**, denoted by ′; thus 28′ means 28 minutes.

(2) Each minute is divided into 60 equal parts, called **seconds**, denoted by ″. For example, 30″ means 30 seconds.

Example.—An angle denoted by 37° 15′ 27″ means 37 degrees, 15 minutes, 27 seconds.

This subdivision of the degree is very important in marine

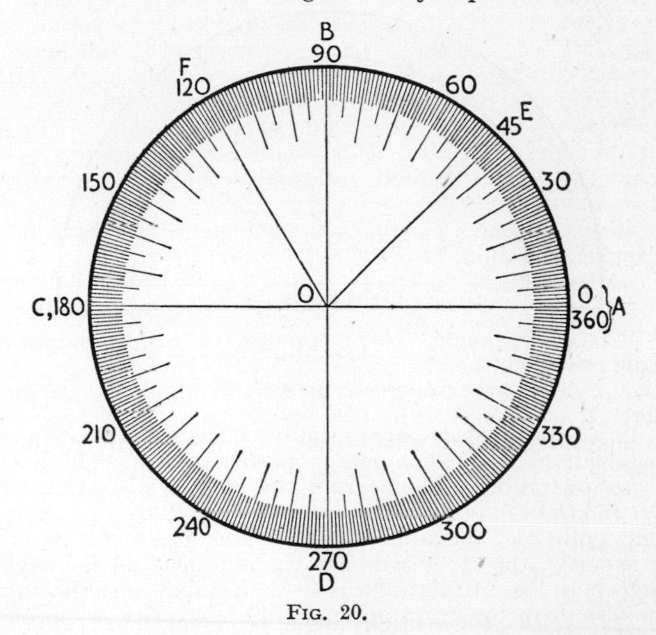

Fig. 20.

and air navigation, surveying, gunnery, etc., where very great accuracy is essential.

It will be observed that the circle in Fig. 20 is divided

by *BD* and *AC* into four equal sectors called **quadrants.** These are numbered the 1st, 2nd, 3rd and 4th quadrants, respectively, *AOB* being the 1st quadrant, *BOC* the 2nd quadrant, etc.

23. Protractors.

An instrument for measuring or constructing angles is called a **protractor.** It may be semi-circular or rectangular in shape. The ordinary semi-circular protractor is much the same as half of the circle shown in Fig. 21.

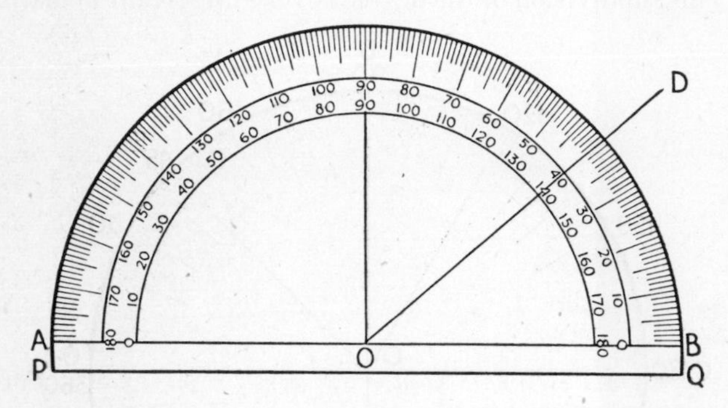

FIG. 21.

A circular protractor.

These protractors are usually made of transparent celluloid, so that when one is placed over straight lines these are visible. To measure the angle whose arms are *OB* and *OD* the protractor is placed with *OB* over one arm. The point on the angle scale of the protractor where it is cut by the other arm enables us to read off the angle *BOD*. In Fig. 21 this angle is 40°. A suitable modification enables us to construct an angle of a given size, when one arm of the angle is fixed.

The purpose of the two sets of numbers is to make it easy to read the angle from either end *A* or *B*. (See the next paragraph.)

24. Supplementary and Complementary Angles.

(1) Supplementary angles.

When the sum of two angles is equal to two right angles, each of the angles is called the supplement of the other.

In Fig. 22 $\angle BOA$ is the supplement of $\angle BOC$,
and $\angle BOC$,, ,, $\angle BOA$.

Example. The supplement of 30° is 180° − 30° = 150°.
Also ,, 150° is 180° − 150° = 30°.

(See scales on protractor, Fig. 21).

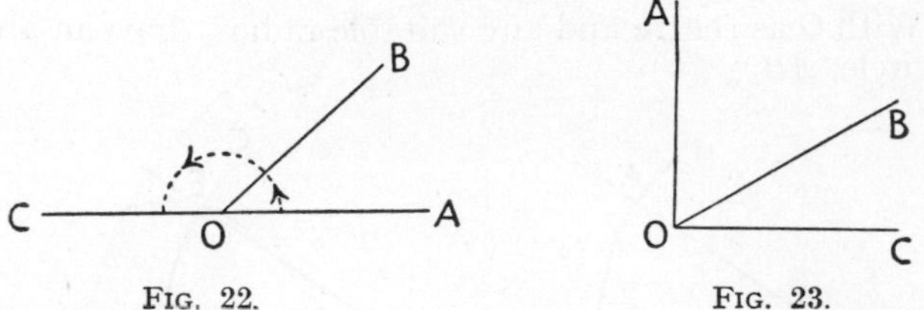

FIG. 22. FIG. 23.

The Theorem of § 18 could therefore be written as follows:

If a straight line meets another straight line, the adjacent angles are supplementary.

(2) Complementary angles.

When the sum of two angles is a right angle each of the angles is the complement of the other.

In Fig. 23 $\angle BOC$ is the complement of $\angle AOB$,
and $\angle AOB$,, ,, $\angle BOC$.

Example. The complement of 30° is 90° − 30° = 60°,
and ,, ,, 60° is 90° − 60° = 30°.

25. A Practical Problem.

The foregoing work enables us to perform a useful piece of practical work, the first of our constructions.

Construction I.

To construct an angle equal to a given angle.

Let $\angle AOB$, Fig. 24, be the angle which we require to copy. We are not concerned with the number of degrees in the angle, and a protractor is not necessary.

Method of Construction.

Take a straight line, PQ, which is to be one of the arms of the required angle.

With O as centre and any suitable radius, draw an arc of a circle, AB.

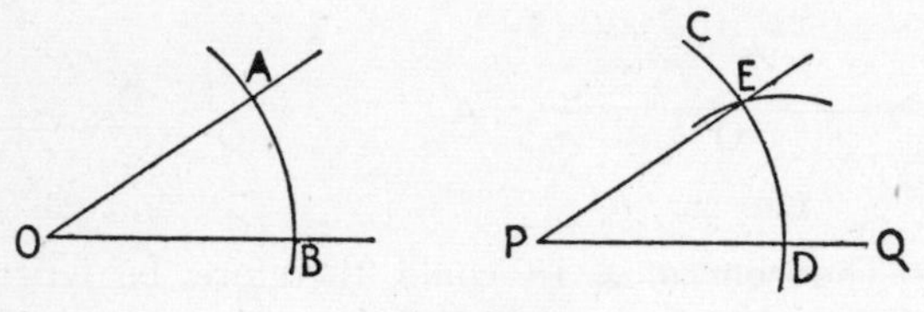

Fig. 24.

With P as centre and the same radius OB, draw another arc, CD.

With D as centre and radius BA, draw another arc intersecting the arc CD in E.

Join EP.

Then $\angle EPD$ is the angle required.

The two circles of which AB and ED are arcs have the same radii. Since DE was made equal to AB, it is evident that the arc ED is equal to the arc AB.

$\therefore$ from previous conclusions the angles at the centre AOB and EPD may reasonably be concluded as equal.

$\therefore$ the angle EPD has been constructed equal to the angle AOB.

Exercise I

1. Using a protractor, measure the angles marked with a cross in Fig. 25, check by finding their sum. What kind of angles are the following?

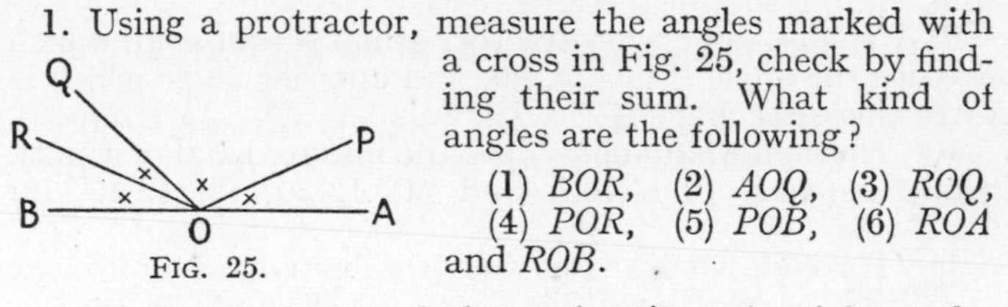

Fig. 25.

(1) BOR, (2) AOQ, (3) ROQ, (4) POR, (5) POB, (6) ROA and ROB.

2. State in degrees and also as fractions of a right angle:

(*a*) the complement, and (*b*) the supplement, of one-fifth of a right angle.

3. (*a*) Write down the complements of $37\frac{1}{2}°$, $45° 15'$, $72° 40'$.
(*b*) Write down the supplements of $112°$, $154° 30'$, $21° 15'$.

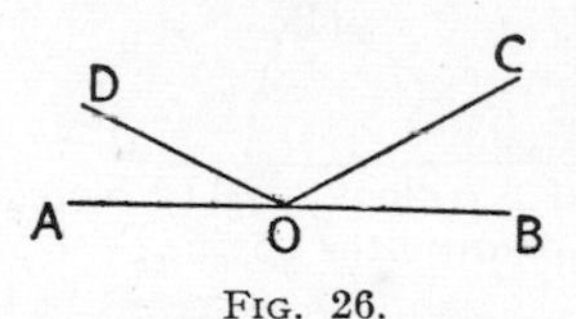

Fig. 26.

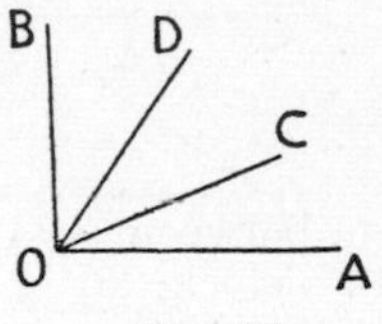

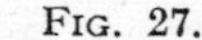
Fig. 27.

4. In Fig. 26 if the $\angle AOD = 25°$ and $\angle BOC = 31°$, Find $\angle COD$. What is its supplement?

5. In Fig. 27 $\angle AOB$ is a right angle and OC and OD are any two straight lines intersecting AO and OB at O. Name the angles which are complementary to AOC, AOD, COB, DOB.

6. In Fig. 28, $\angle AOB$ is an acute angle and OP, OQ are drawn perpendicular to OA and OB, respectively. What reason could you give to justify the statement that $\angle AOB = \angle POQ$?

7. Without using a protractor, construct angles equal to A and B in Fig. 29. Afterwards check by measuring the angles with a protractor.

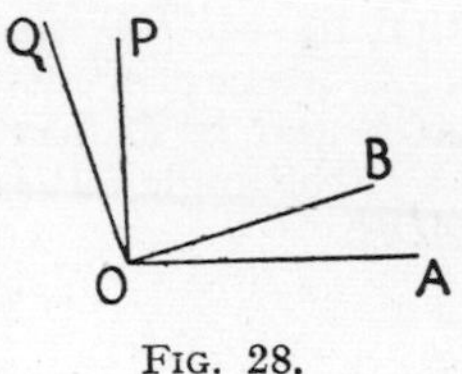

Fig. 28.

8. Draw a straight line, PQ. At P on one side of it, construct an angle of $72°$. On the other side con-

struct an angle of 28°. Check by measuring the angle which is the sum of these.

9. Without using a protractor, construct an angle which is twice the angle B in Fig. 29, and another angle which is three times the angle A.

10. Through what angles does the minute hand of a clock rotate between 12 o'clock and (1) 12.20, (2) 12.45, (3) 2 o'clock?

11. Through what angles does the hour hand of a clock

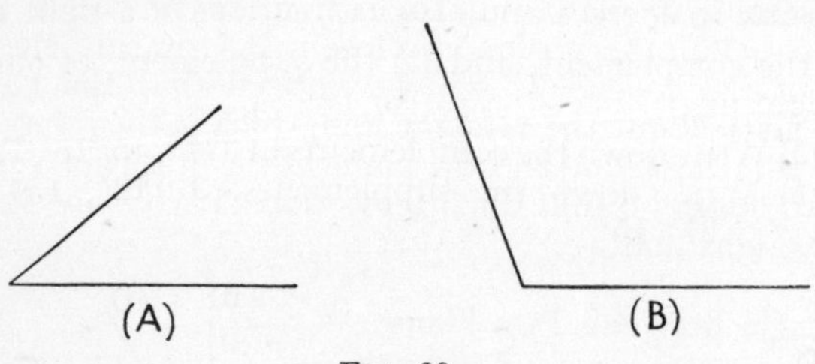

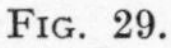
Fig. 29.

rotate between (1) 12 o'clock and 2 o'clock, (2) 12 o'clock and 6 o'clock, (3) 12 o'clock and 10 minutes to one?

CHAPTER 4

SIMPLE GEOMETRY OF PLANES

26. Rotation of a Plane.

Every time that you turn over a page of this book you are rotating a plane surface, or, more briefly, a plane: this may be observed more closely by rotating the front page of the cover. It will be noticed that **the rotation takes place about the straight line** which is the intersection of the rotating plane and the plane of the first page. It was pointed out in § 3 that the intersection of two plane surfaces is a straight line.

27. Angle between Two Planes.

Take a piece of fairly stout paper and fold it in two. Let AB, Fig. 30, be the line of the fold. Draw this straight line. Let $BCDA$, $BEFA$ represent the two parts of the paper.

These can be regarded as two separate planes. Starting with the two parts folded together, keeping one part fixed, the other part can be rotated about AB into the position indicated by $ABEF$. In this process the plane $ABEF$ has moved through an angle relative to the fixed plane. This is analogous to that of the rotation of a line as described in § 16. We must now consider how this angle can be definitely fixed and measured. Flattening out the whole paper again, take any point P on the line of the fold—*i.e.*, AB, and draw RPQ at right angles to AB. If you fold again, PR will

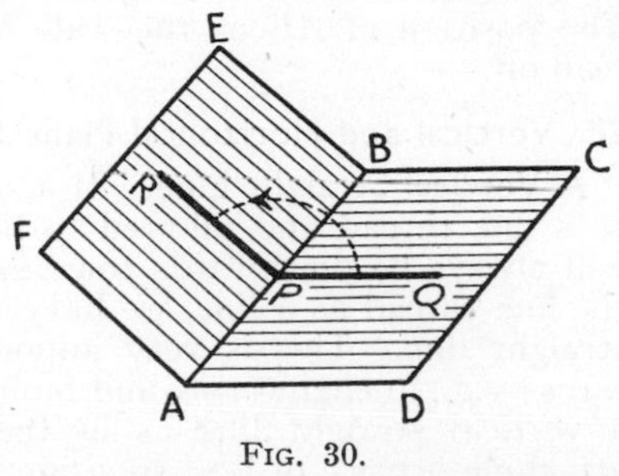

FIG. 30.

coincide with PQ. Now rotate again, and the line PR will mark out an angle RPQ relative to PQ. The angle RPQ is thus the angle which measures the amount of rotation, and is called the angle between the planes.

Definition. *The angle between two planes is the angle between two straight lines which are drawn, one in each plane, at right angles to the line of intersection of the plane and from the same point on it.*

When this angle becomes a right angle the **planes are perpendicular to one another.**

Numerous examples of planes which are perpendicular to each other may be observed. The walls of the room are perpendicular to the floor and ceiling; the surfaces of the cover of a match-box, as shown in Fig. 2, are perpendicular to each other, when they intersect.

The angle between two planes can be measured, in many cases, by means of a protractor. If, for example, it is required to measure the angle between the two planes in Fig. 30, this can be done by measuring the angle DAF, provided that AF and AD are perpendicular to the axis AB. The protractor is placed so that the point O, Fig. 21, is at A and the line AD coincides with OB in the protractor. The position of AF on the scale of the protractor can be read off.

28. Vertical and Horizontal Plane Surfaces.

A Vertical Straight Line. If a small weight be attached to a fine thread and allowed to hang freely, its direction will always be downwards towards the earth. Regarding the fine thread as a line, we have what is called a **vertical straight line.** This is very important in many practical ways—*e.g.*, in engineering and building. A builder obtains a vertical straight line as in the above experiment by attaching a piece of lead to a fine string. This is called a **plumb line.**

A Vertical Surface. A plane surface which contains two or more vertical straight lines is a **vertical surface.** For example, the surfaces of walls of a house are vertical.

Horizontal Surface. Take a glass containing water and suspend in it a weight held by a thread (Fig. 31). Let a couple of matches float on the surface, so that ends of the matches touch the string. It will be observed that the string is always at right angles to the matches. In other words, the string is always perpendicular to any straight lines which it intersects on the surface. Under these conditions the **thread is said to be perpendicular to the surface.**

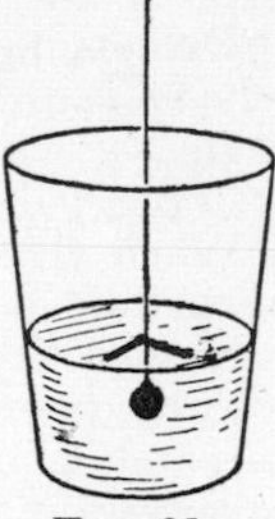

FIG. 31.

A surface which is thus at right angles to a vertical line or surface is called a **horizontal surface.** Or it may be stated thus: **a horizontal surface is always perpendicular to vertical lines or surfaces which intersect it.**

29. A Straight Line Perpendicular to a Plane.

The experiment above leads to a general consideration of the conditions under which a straight line is perpendicular to a plane.

Take a piece of cardboard, *AB* (Fig. 32), and on it draw a

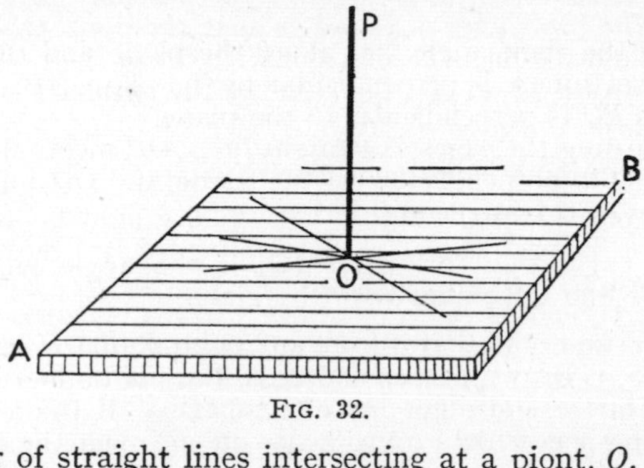

FIG. 32.

number of straight lines intersecting at a piont, *O*. At *O* fix a pin, *OP*, so that it is perpendicular to one of these lines. Then *OP* will be perpendicular to the other lines and is said to be perpendicular to the plane *AB*.

Definition. *A straight line is said to be perpendicular to a plane when it is perpendicular to any straight line which it meets in the plane.*

The distance of the point P from the plane AB is given by the length of the perpendicular, OP, drawn from it to the plane.

30. Angle between a Straight Line and a Plane which it Meets.

Take a set square, *OPQ*, and stand it on a piece of smooth paper or cardboard, *AB*, so that one of the edges, *OQ*, con-

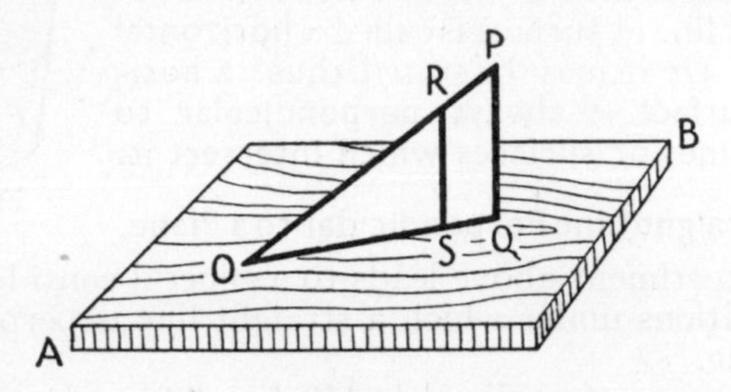

FIG. 33.

taining the right angle, lies along the plane and the plane of the set square is perpendicular to the plane *AB*.

Thus *PQ* is perpendicular to the plane.

Regarding the edges as straight lines, *OP* meets the plane in *O*, *PQ* is perpendicular to the plane and *OQ* joins *O* to the foot of this perpendicular.

The angle thus formed, POQ, is the angle which the straight line OP makes with the plane.

It will be noted that if from **any** point *R* on *OP* a straight line, *RS*, is drawn perpendicular to the plane, *S* will lie on *OQ*. Thus the straight line *OQ* contains all the points in which perpendiculars from points on *OP* meet the plane.

OQ is called the projection of OP on the plane AB.

Definition. *The angle between a straight line and a plane is the angle between the straight line and its projection on the plane.*

Consequently, the projection of a straight line OP *on a*

plane which it meets at O, *is the straight line intercepted on the plane between* O *and the foot of the perpendicular drawn from* P *to the plane.*

This may be extended to the case in which the straight line does not meet the plane. Thus in Fig. 33, QS is the projection of PR on the plane AB.

CHAPTER 5

DIRECTION

31. Meaning of Direction.

The term " direction " is a difficult one to define, but its meaning is generally understood, and the definition will not be attempted here. It is, however, often used vaguely, as when we speak of walking " in the direction of London ". We are more precise when we speak of the direction of the wind as being, say, " north-west ", though this may sometimes be only roughly correct. To find exact direction is so important in navigation, both at sea and in the air, as well as in many other ways, that it is desirable to have precise ideas of what is understood by " direction " and how it is

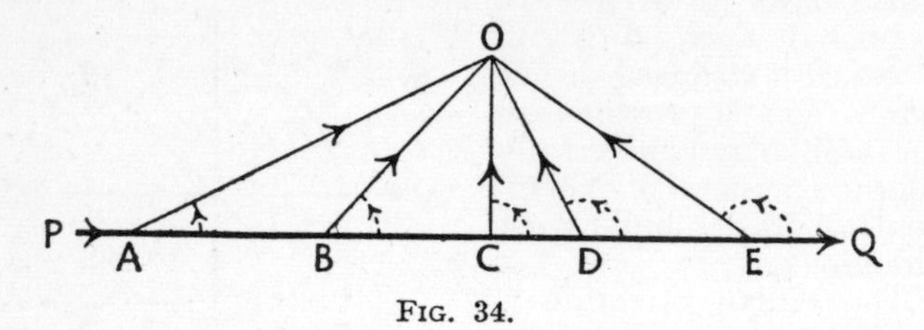

FIG. 34.

determined and expressed. We will begin with a simple everyday example.

In Fig. 34, PQ represents a straight road along which a man walks from P towards Q. O represents the position of a church tower lying at some distance from the road. At various points along the road, A, B, C, D, E, the straight lines AO, BO, CO, DO, EO represent the direction of O at these points. This direction can be described more accurately if we know the angle which the line of direction makes with the road. The angles, of course, change as the man walks along, as is evident from the diagram, in which the angles are consistently measured in an anti-clockwise direction from the road. These angles can be obtained by the use of a surveying instrument known as a theodolite. If, as an example, the angle made by *BO* with the road is 45°, then we can say that **at B the direction of O makes an angle of 45° with the direction of the road.**

It must be emphasised that this statement as to the direction of *O* gives the information **only relative to the direction of the road,** and this may not be known. Consequently for practical purposes the statement is not precise and does not state an absolution direction.

32. Standard Direction.

All directions are relative—*i.e.*, they are related to some other direction, as in the case of the road above. Therefore it is necessary for practical purposes—*e.g.*, navigation, that there should be some selected fixed direction to which other directions can be related. Such a direction is called a **standard direction.** This is provided by the familiar and universally adopted system of North, South, East and West directions.

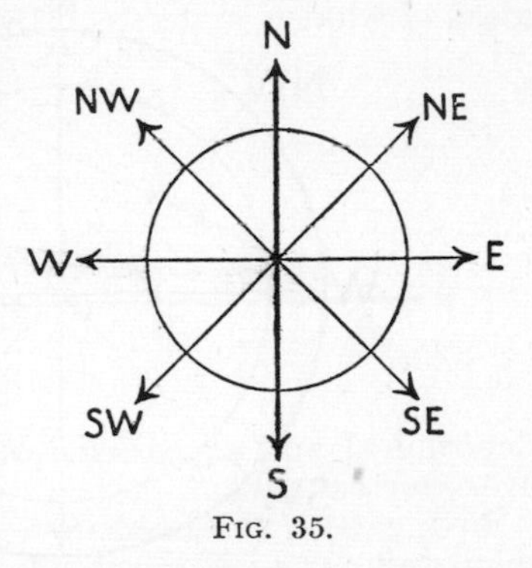

FIG. 35.

The **North direction** is fixed by the position of the **North Pole,** which is an imaginary point at the end of the Earth's axis (see § 189 and Fig. 182). The **South** is the opposite direction from the North. **East** and **West** directions are at right angles to these.

These four directions are termed the **cardinal points.** They are indicated in Fig. 35 and all others between them are related to these. Thus a direction half-way between N. and E., and thus making angles of 45° with each, is called North-east, and so for others as shown in Fig. 35.

33. The Magnetic Compass.

The North direction can always be determined by the use of the **Magnetic compass** or **Mariners' compass.** This instrument has a magnetised needle which is free to move in a horizontal plane: the needle always sets, not towards the true North, but in the direction of what is called the **Magnetic North.** The amount of the angle of deflection from the true North is known at various positions on the Earth's surface, and with this correction the true North can readily be found.

34. Points of the Compass.

Fig. 36 illustrates part of the compass card or dial of a mariners' compass. Two diameters at right angles to one

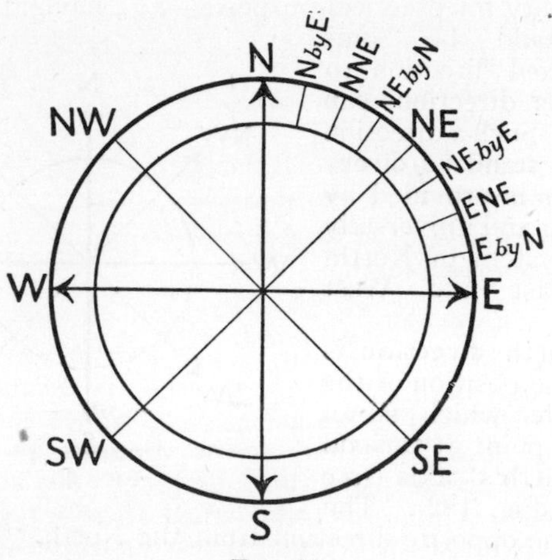

Fig. 36.

another, and representing North–South and East–West, divide the circle into four quadrants. Each of the quadrants is further subdivided into eight equal divisions. Thus the whole circle has thirty-two divisions each of which represents a definite direction. The names employed to indicate these directions are shown for the first quadrant; those in other quadrants are similarly divided and described. The arc of each of these thirty-two divisions subtends an angle at the centre which must be $(\frac{360}{32})^\circ$ or $11\frac{1}{4}^\circ$.

Directions between these are indicated by stating the number of degrees, from one of the 32 fixed directions, thus:

6° East of North indicates a point between N. and N. by E., and 6° from the North.

35. Bearing.

When the direction of one object, B, with respect to another object A is defined by reference to a standard direction, the angle giving this direction is called the bearing of B from A.

Thus, in Fig. 37, if A and B represent two ships, and the angle BAN gives the direction of B from the North, then:

The angle BAN is called the bearing of B with respect to A.

If the angle BAN is 40°, then the bearing of B from A is 40° East of North.

Bearings are measured in a clockwise direction from the North.

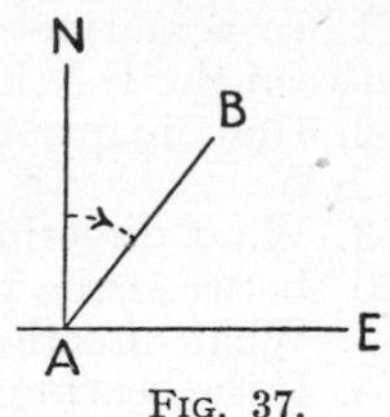

FIG. 37.

36. Angle of Elevation.

In the consideration of direction we have so far been concerned only with direction on the horizontal plane. But if an object such as an aeroplane is above the surface of the Earth, in order to find its true direction its position **above the horizontal plane** must be taken into account.

If, for example, a gun is to be aimed at an aeroplane, we must not only know the horizontal bearing of the aeroplane, but we must also know the angle through which it is necessary to elevate the gun to point to it. This angle is called the **angle of elevation** of the aeroplane.

In Fig. 38, if A represents the position of the aeroplane, and O the position of the gun, then the latter must be rotated from the horizontal in a vertical plane through the angle AOB to point to the aeroplane.

The angle AOB is called the angle of elevation of the aeroplane.

The determination of the **actual height** of the aeroplane when the angle of elevation is known, is a problem which requires Trigonometry for its solution.

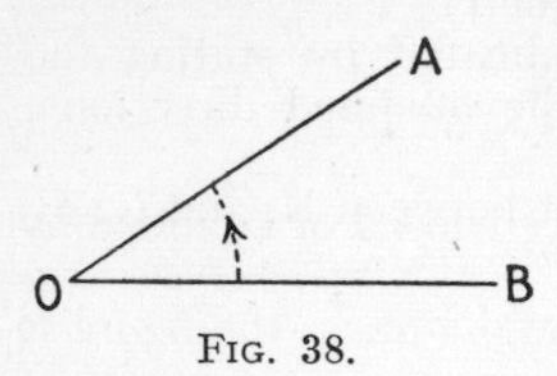

Fig. 38.

37. The Altitude of the Sun.

The angle of elevation of the sun is called its altitude. The determination of the altitude of the sun is of great importance in marine navigation. The instrument which is used for the purpose is called a **sextant.**

Exercise 2

1. In a mariners' compass how many degrees are there between the E.N.E. and N.N.E. directions?

2. How many degrees are there between S.S.E. and W.S.W.?

3. What direction is exactly opposite to E. by S.? (*i.e.*, 180° between the two directions).

4. What direction is exactly opposite to E.N.E.?

5. If the bearing of an object is E.S.E., how many degrees is this?

6. A ship sailing N.N.E. changes its course by turning through an angle of $67\frac{1}{2}°$. What is then the direction of its course?

7. If the direction of an aeroplane makes an angle of 57° 20′ with the horizontal plane, what angle does it make with the vertical plane?

8. Two straight lines, *AB* and *AD*, lie in the same vertical plane. *AB* makes an angle of 25° with the horizontal, and *AD* makes 32° with the vertical. What is the angle between *AB* and *AD*?

CHAPTER 6

TRIANGLES

38. Rectilineal Figures.

A part of a plane surface which is enclosed or bounded by lines is called a **plane figure.**

If the boundary lines are all straight lines, the figure is called a **rectilineal figure.**

The least number of straight lines which can thus enclose a space is **three.** It was stated in axiom 3, § 6, that two straight lines cannot enclose a space.

When **three** straight lines intersect, the part of the plane

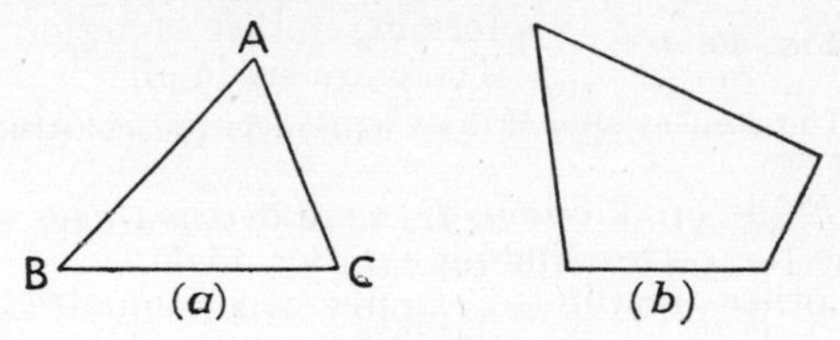

FIG. 39.

enclosed is a triangle, as Fig. 39 (*a*). Three angles are formed by the intersection of the straight lines; hence the name.

When four straight lines intersect in the same plane, the figure formed is a **quadrilateral** (Fig. 39 (*b*)).

Perimeter. **The sum of the lengths of the sides of a rectilineal figure is called its perimeter.** Thus, for the triangle in Fig. 39 (*a*).

$$\text{Perimeter} = AB + BC + CA.$$

Area. The amount of the surface enclosed by the sides of a rectilineal figure is called its **area.**

39. The Triangle.

Vertex. Each of the angular points of a triangle, as A, B and C in Fig. 39 (*a*), is called a vertex (plural vertices).

Base. When anyone of the three angular points of a triangle is regarded as a vertex, **the side opposite to it is called a base,** or, more accurately, **the corresponding base.**

In Fig. 39 (*a*) if *A* be regarded as a vertex, then *BC* is the corresponding base.

40. Exterior Angles.

If a side of a triangle be produced, the angle so formed with the adjacent side is called an exterior angle.

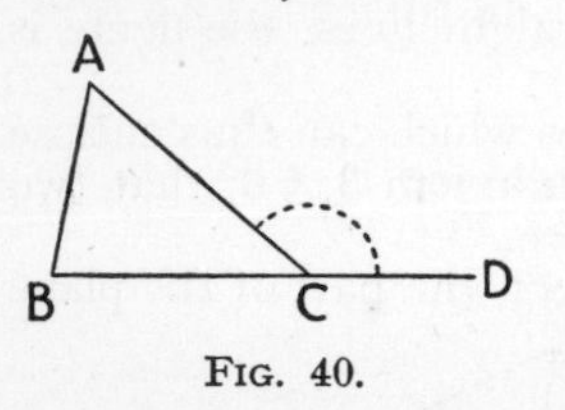

FIG. 40.

In Fig. 40 the side *BC* is produced to *D*, thus forming with the adjacent side, *AC*, the angle *ACD*. This is an exterior angle. Similarly, each of the sides can be produced in two directions, thus forming other exterior angles. There are six in all.

Note.—The student should draw a triangle and construct all the exterior angles.

When *BC* is produced to *D*, as above, we may speak of $\angle ACD$ as the corresponding exterior angle.

With other rectilineal figures—*e.g.*, quadrilaterals—exterior angles may similarly be constructed.

41. Kinds of Triangles.

Triangles may be classified: (1) according to their angles, and (2) according to their sides.

(1) Triangles classified according to angles.

(*a*) A triangle having **one of its angles obtuse** is called an **obtuse-angled triangle** (Fig. 41 (*a*)).

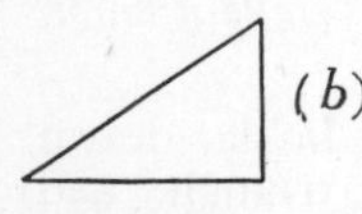

(*b*) When **one of the angles is a right** angle, the triangle is a **right-angled tri**angle (Fig. 41 (*b*)).

FIG. 41.

The side opposite to the right angle is called the **hypotenuse.**

A triangle with **all its angles acute** is called an **acute-angled triangle** (Fig. 41 (*c*)).

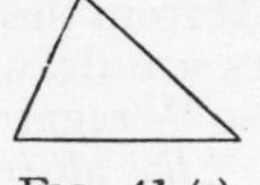

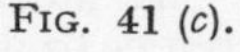

FIG. 41 (*c*).

(2) **Triangles classified according to sides.**

(*a*) **A triangle with two equal sides is called isosceles,** as in Fig. 42 (*a*). The angular point between the equal sides is called the **vertex** and the side opposite to it the **base.**

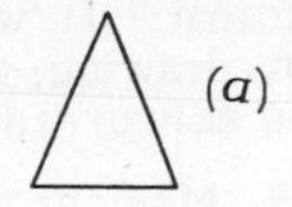

(*b*) When all the sides of the triangle are equal, the triangle is **equilateral,** as Fig. 42 (*b*).

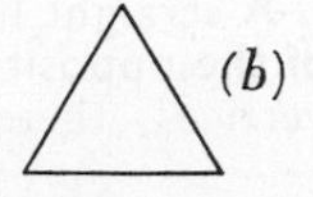

(*c*) When all the sides are unequal, the triangle is called a **scalene triangle** (Fig. 42 (*c*)).

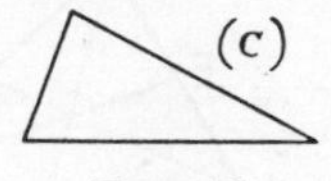

FIG. 42.

42. Altitude of a Triangle.

In triangle *ABC* (Fig. 43 (*a*)), let *A* be regarded as a vertex and *BC* as the corresponding base.

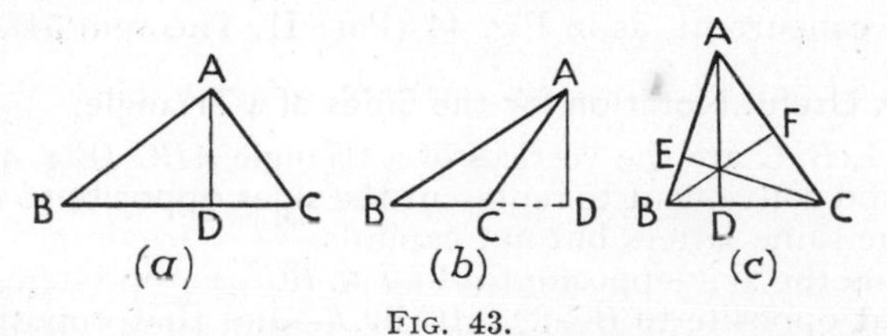

FIG. 43.

From *A* draw *AD* perpendicular to *BC*.

AD is called an altitude (or height) of the triangle when A is the vertex.

If the triangle is obtuse angled, the perpendicular drawn from *A*, as in Fig. 43 (*b*), falls outside the triangle, and the basc *BC* must be produced to meet it.

If from the other vertices *B* and *C* (Fig. 43 (*c*)) perpendiculars are drawn to corresponding opposite sides, these may also be regarded as altitudes. Consequently when speaking of "*the altitude of a triangle*" it must be understood that it refers to a particular vertex and the corresponding base. In Fig. 43 (*c*) three altitudes of the $\triangle ABC$ have been drawn. It will be proved later, and it may be verified by drawing, that they meet in a point. They are said to be **concurrent** (Part II, Theorem 55).

43. Medians of a Triangle.

A straight line which joins a vertex to the middle point of the opposite side is called a median. As there are three vertices, there are three medians. In Fig. 44, *AD*, *BF* and *CE* are medians. These medians can also be shown to be concurrent, as in Fig. 44 (Part II, Theorem 54).

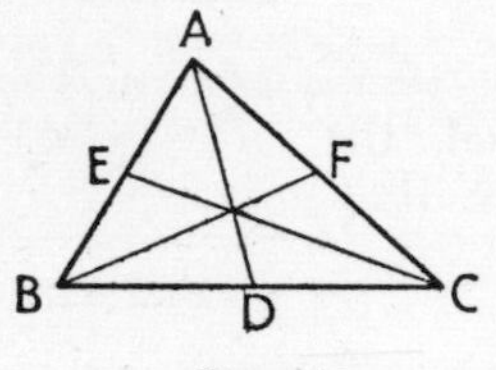

FIG. 44.

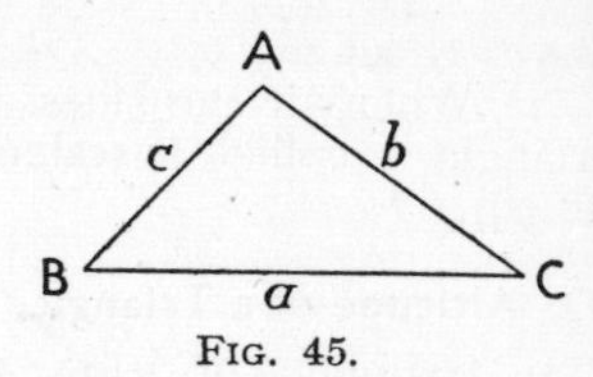

FIG. 45.

44. A Useful Notation for the Sides of a Triangle.

If *A*, *B*, *C* are the vertices of a triangle *ABC* (Fig. 45), it is found convenient to represent the **sides opposite to them** by the same letters but not capitals.

Thus the side opposite to *A*—*i.e.*, *BC*—is represented by *a*, that opposite to *B*—*i.e.*, *AC* by *b*—and that opposite to *C* by *c*. This notation makes easier the identification of corresponding sides and angles.

Congruent triangles

45. Construction of Triangles from Fixed Data.

The method of constructing a triangle varies according to the facts which are known about the sides and angles.

We shall proceed to discover what is the **minimum** knowledge about the sides and angles which is necessary to construct a particular triangle. It will be found that if any one of **three different sets** of equal angles or sides is known the triangle can be constructed. These sets are **A**, **B** and **C** below.

A. Given the lengths of two sides and the size of the angle between them.

Example. *Construct the triangle in which the lengths of two sides are 2 in. and 1·5 in., and the angle between them is 40°.*

I.e., Given, two sides and the angle between them.

Note.—The student should himself take a piece of smooth paper and carry out, step by step, the method of construction given below.

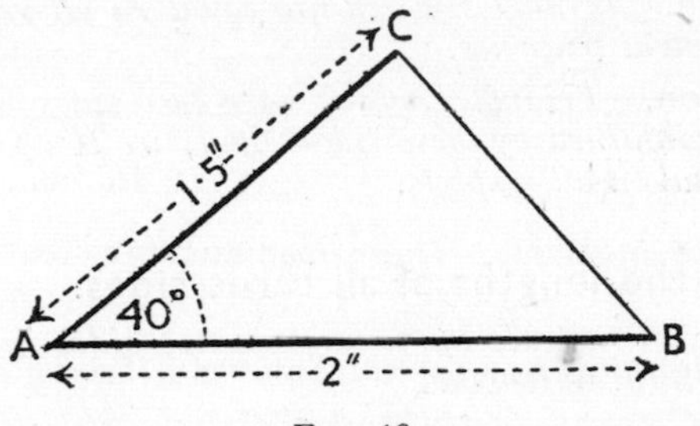

Fig. 46.

Construction.

(1) Draw the straight line *AB*, 2 in. long (Fig. 46).

(2) At *A*, with a protractor, draw a line making an angle of 40° with *AB*.

(3) From this line cut off *AC* 1·5 in. long.

Thus the points **B and C are fixed points,** and they must be vertices of the required triangle. Join *BC*. This must be the third side of the triangle, and *ABC* must be the triangle required.

Since only one straight line can be drawn through the two points B and C (axiom 1, § 6), the triangle can be completed in one way only, and is therefore drawn as in Fig. 46.

Thus, BC and the angles ACB and ABC are fixed, and there can be only one triangle which has the sides and angle of the given dimensions.

The student should cut out the $\triangle$ he has constructed and place it over that in Fig. 46. If the reasoning has been correct and the drawing accurate, the two $\triangle$s should exactly coincide.

It is evident that if all the students who read this book were to construct $\triangle$s with sides and angle as above, all the $\triangle$s would be of exactly the same size and shape—*i.e.*, they will coincide and their areas must be the same.

Triangles which are equal in every respect and coincide in this way are said to be congruent.

Definition. *Triangles which are equal in all respects are called congruent triangles.*

Conclusion. *Triangles which have two sides equal, and the angles contained by these sides equal, are congruent*—i.e., *they are equal in all respects.*

B. Given the lengths of all three sides.

Example. *Construct the triangle whose sides are* 2·5 *in.,* 2 *in. and* 1·5 *in. in length.*

Construction.

(1) Draw a straight line AB, 2·5 in. long (Fig. 47).

(2) With centre B and radius 1·5 in. draw an arc of a circle.

(3) With centre A and radius 2 in. draw an arc of a circle.

The two arcs cut at C.

$\therefore$ C is 2 in. from A and 1·5 in. from B.

Joining AC and CB the $\triangle ABC$ so formed fulfils the given conditions.

Clearly there can be one triangle only, and all $\triangle$s constructured as above must be identically equal. ∴ they must be congruent. As before the student should construct a triangle step by step, and test his △ by the one in Fig. 47.

Note.—If the complete circles be drawn above, with A and B as centres, they will cut at a second point on the other side of AB. The △ so obtained is clearly identical with △ACB. (See Fig. 116.)

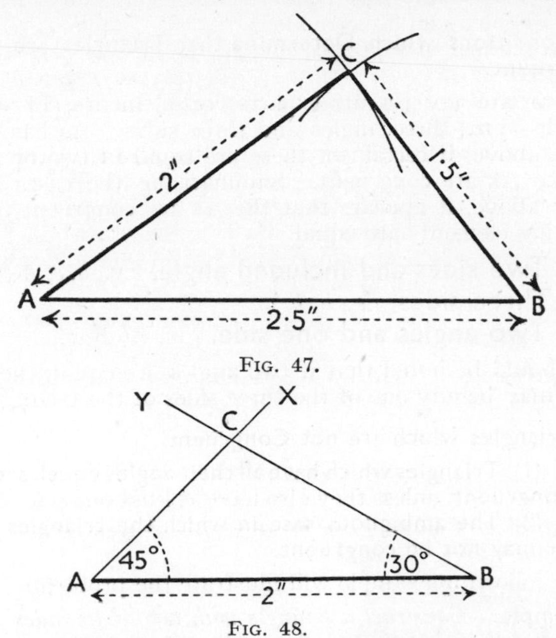

FIG. 47.

FIG. 48.

C. Given two angles of the triangle and one side.

Example. *Construct a triangle having one side* 2 *in. long and two angles* 30° *and* 45°.

Construction.

(1) Draw a straight line AB, 2 in. long (Fig. 48).
(2) At A draw AX making 45° with AB.
(3) At B „ BY „ 30° „ AB.

The two straight lines AX and BY will cut at one point only, C.

Thus ACB is a triangle which fulfils the conditions and is the only one possible.

$\therefore$ all other such Δs drawn in the same way with the same data will be congruent.

46. Conditions which Determine that Triangles are Congruent.

There are six essential parts, or elements, of every triangle—viz., three angles and three sides. As has been shown above, if certain of these are equal in two or more Δs, the Δs are congruent. Summarising the cases A, B and C above, it appears that the Δs are congruent if the following elements are equal.

A. Two sides and included angle, *e.g.*, b, c, A (§ 44).
B. Three sides, *i.e.*, a, b, c.
C. Two angles and one side, *e.g.*, A, B, c.

It should be noted that if two angles are equal, the side given may be any one of the three sides of the triangle.

47. Triangles which are not Congruent.

(1) **Triangles which have all their angles equal, are not congruent,** unless they also have at least one side equal.

(2) **The ambiguous case in which the triangles may or may not be congruent.**

The following example will illustrate the problem.

Example. *Construct a triangle with two of its sides* 2 *in. and* 1·2 *in. in length and the angle opposite to the smaller side* 30°.

Construction.

(1) Draw a straight line AD of indefinite length (Fig. 49).

(2) At a point A draw a straight line making an angle of 30° with AD.

(3) From this line cut off AC 2 in. long.

(4) With C as centre and radius 1·2 in. draw an arc

of a circle which will cut AD. This it will do in two points, B and B'.

(5) Join CB and CB'.

Thus **two triangles are constructed ACB and ACB′ each of which satisfies the given conditions.**

Thus the solution is **not unique** as in cases A, B, C above —*i.e.*, there is not **one** triangle, satisfying the given conditions, which may be drawn as in previous cases. **There can be two triangles.**

Hence this is called the **ambiguous case.**

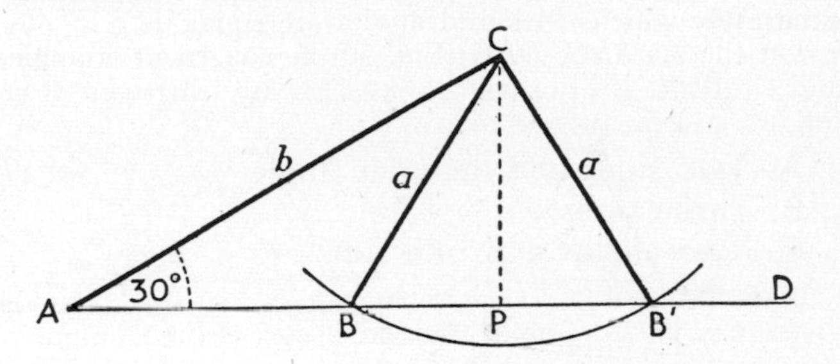

FIG. 49.

There are, however, two cases in which there is no ambiguity. Using the notation of § 44, in Fig. 49 it will be seen that we had given above **A, a, b**—*i.e.*, **two sides and an angle opposite to one of them.**

From C (Fig. 49) draw CP perpendicular to AD.

Let $CP = h$.

(1) **If a = h there is one solution** and one only—viz., $\triangle ACP$. There is thus no ambiguity. The triangle is right angled.

(2) **If a = b** there **will be one solution,** and the triangle will be isosceles, the other side, equal to CA meeting AB produced.

(3) If $a > b$ **there will be one solution** as is obvious. It will be seen therefore that **for ambiguity a must be less than b and greater than h.**

Summarising.

If the given elements are A, a, b, ambiguity will arise if a, *i.e.*, the side opposite to the given angle is less than b, unless a is equal to the perpendicular drawn from C to the side c.

48. Corresponding Sides and Angles of Congruent Triangles.

When triangles are congruent it is important to specify accurately which sides and angles are equal.

Let the $\triangle$s ABC, DEF (Fig. 50) be congruent triangles.

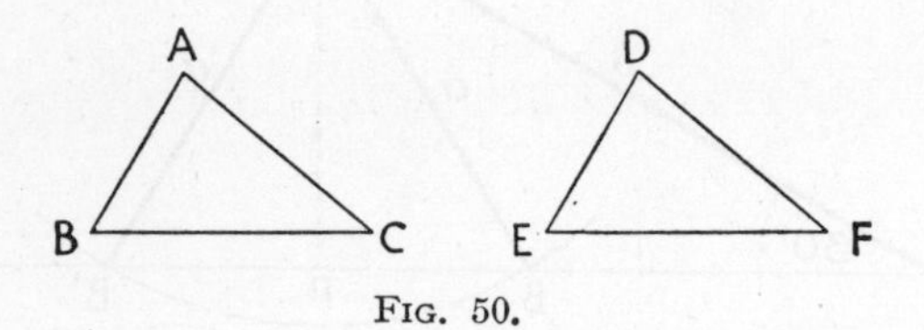

FIG. 50.

If AB and DE are sides which are known to be equal, then **the angles which are opposite to them are called corresponding angles and are equal.**

Summarising:

If $AB = DE$, $\angle ACB = \angle DFE$
$BC = EF$, $\angle BAC = \angle EDF$
$CA = FD$, $\angle ABC = \angle DEF$.

Similarly when angles are known to be equal, the opposite sides are **corresponding sides** and are equal.

49. Theorems concerning Congruent Triangles.

The three sets of conditions that triangles may be congruent, which were deduced in § 45, may be set out in the form of Geometric Theorems as follows:

Theorem A. Two triangles are congruent if two sides and the included angle of one triangle are respectively equal to two sides and the included angle of the other.

Theorem B. Two triangles are congruent if the three sides of one triangle are respectively equal to the three sides of the other.

Theorem C. Two triangles are congruent if two angles and a side of one triangle are respectively equal to two angles and a side of the other.

It will be noted that the theorems above have been enunciated, or stated, with respect to **two** triangles, because it is in this form that the theorem is usually applied, but they are true, of course, for **all** triangles which satisfy the given conditions.

Exercise 3

1. Construct a triangle in which two of the sides are 3 in. and 4 in. and the angle between them 35°. Find by measurement the third side and the other angles.
2. Construct a triangle of which the three sides are 5 in., 5·5 in. and 6 in. Measure the angles and find their sum.
3. Construct a triangle in which two of the angles are 40° and 50° and the length of the side adjacent to them both is 3 in. Measure the third angle and the lengths of the other two sides.
4. The angles and sides of a triangle are as follows.

$$A = 88°, B = 40°, C = 52°.$$
$$a = 6{\cdot}15 \text{ in.}, b = 3{\cdot}95 \text{ in.}, c = 4{\cdot}85 \text{ in.}$$

Construct the triangle in three different ways by selecting appropriate data. Cut out the triangles and compare them.

5. Construct by three different methods triangles congruent with that in Fig. 51. Cut them out and test by superimposing them on one another.

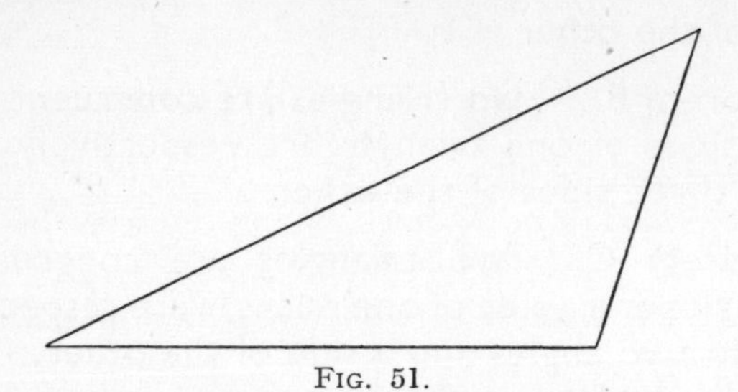

FIG. 51.

6. In the $\triangle$s ABC, DEF certain angles and sides are equal as given below. Determine whether or not they are congruent. If they are congruent, state which of the conditions A, B, C of § 49 is satisfied.

(1) $\angle A = \angle D,\ b = e,\ c = f.$
(2) $\angle A = \angle D,\ c = f,\ a = d.$
(3) $\angle B = \angle E,\ b = e,\ c = f.$
(4) $\angle B = \angle E,\ \angle A = \angle D,\ c = f.$
(5) $\angle D = \angle E,\ \angle A = \angle D,\ b = e.$

7. In a $\triangle ABC$, $\angle B = 25°$, $c = 4{\cdot}35$, $b = 3{\cdot}8$. Construct the triangle and show that there are two solutions.

8. Two straight lines AB, CD bisect each other at their point of intersection O. What reasons can you give for saying that $CB = AD$?

9. The straight lines AB and CD bisect each other at O. What reasons can you give for stating that the straight lines AC, CB, BD, DA are all equal?

10. The $\triangle$s ACB, ADB are congruent and are placed on opposite sides of the common side AB. Join CD cutting AB at O. Using Theorem A show that $OC = OD$.

CHAPTER 7

PARALLEL STRAIGHT LINES

50. Meaning of Parallel.

If the ruled printed lines on an exercise book are examined, two facts will be evident.

(1) The distance between any pair of lines is always the same.

(2) If the lines could be produced through any distance beyond the page of the book, you would be confident that they would never meet.

Such straight lines drawn in a plane are called parallel straight lines.

Railway lines provide another example of parallel straight lines. We know that they must be always the same distance apart, and no matter how far they extend they will never meet. If they did it would be a bad business for a train travelling over them.

All vertical straight lines are parallel. If weights be allowed to hang freely at the end of threads, the threads are always parallel.

In Geometry parallel straight lines are very important, and we need to be thoroughly acquainted with certain geometrical facts about them. They are defined as follows:

Definition. *Parallel straight lines are such that lying in the same plane, they do not meet, however far they may be produced in either direction.*

51. Distance between Two Parallel Straight Lines.

It should be noted that in the definition of parallel straight lines stated above there is no mention of them being always the same distance apart, though in the preceding explanations it was stated as a fact which would be regarded as obvious. The definition itself involves only the fact **that parallel straight lines in a plane never meet.** The fact that the distance between them is constant follows from the definition, as will be seen.

It is important that we should be clear at the outset as to what exactly is the "**distance apart**" of parallel straight lines and how it can be measured.

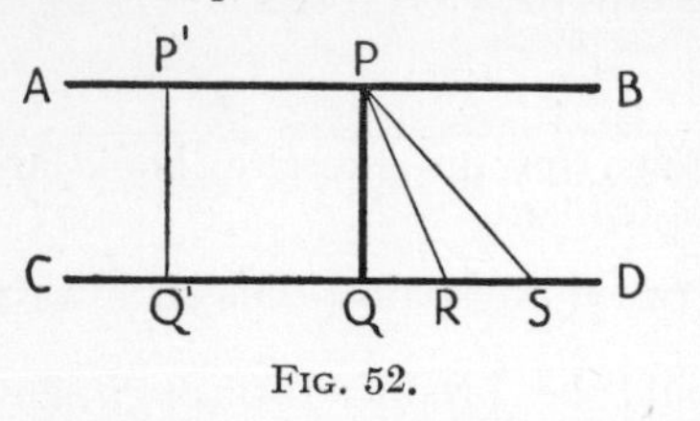

FIG. 52.

Let AB, CD be two parallel straight lines (Fig. 52).

Let P be any point on AB.

Let PQ be the straight line which is the perpendicular drawn from P to CD.

PQ is defined as the distance between the two parallel straight lines. Two facts may be deduced from this:

(1) PQ is the least of all straight lines such as PQ, PR, PS . . . which may be drawn from P to meet CD.

(2) If from any other point P', the straight line $P'Q'$ be drawn perpendicular to CD, $P'Q'$ will also be the distance between the two parallel straight lines, and $\therefore P'Q' = PQ$.

It will be seen later that PQ and $P'Q'$ must themselves be parallel. Consequently we may deduce the fact that: *Straight lines which are perpendicular to parallel straight lines are themselves parallel.*

52. Corresponding Angles.

Take a set square, angles 60°, 30°, 90°, and place the shortest side AC against a ruler, as in Fig. 53. Draw straight lines along the sides of the set square, so forming the triangle ABC.

Now, holding the ruler firmly, slide the set square along it to a new position to form another triangle $A'B'C'$.

The two triangles ABC and $A'B'C'$ must be congruent. The angles at A are right angles and

$$\angle ACB = \angle A'C'B' = 60°.$$

(1) **The straight lines AB and A′B′, which are both perpendicular to the straight line representing the edge of the ruler PQ, are evidently parallel.**

We cannot prove that they satisfy the definition of parallel lines—viz., that they do not meet if produced in

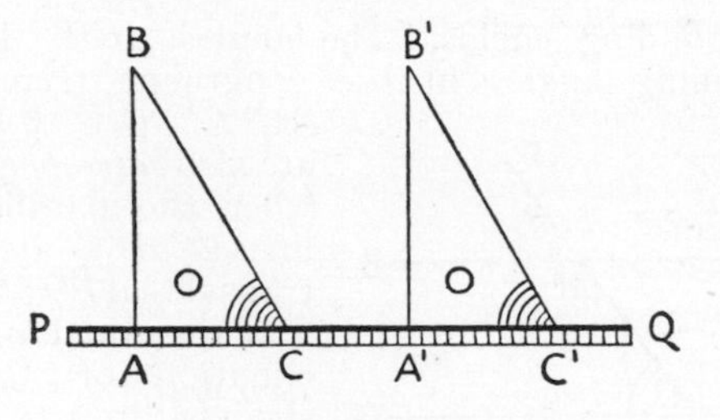

FIG. 53.

either direction—but we know intuitively that they will not meet.

If the set squares in the experiment were moved along the ruler to other similar positions the straight lines corresponding to AB would all be parallel.

This, in effect, is the method commonly employed by draughtsmen for drawing parallel straight lines.

(2) **The angle ACB represents the inclination of the straight line BC to PQ.** Or we may say that BC is inclined at an angle represented by BCP to PQ—*i.e.*, **inclined at 60° to PQ.**

Similarly $\angle A'C'B'$ represents the inclination of $C'B'$ to PQ, and is equal to $\angle ACB$.

$\therefore$ BC and $B'C'$ are equally inclined to PQ.

Also BC and $B'C'$ are evidently parallel straight lines, since, as was the case with AB and $A'B'$, they will obviously never meet if produced. It may now reasonably be deduced that:

(*a*) **Straight lines which are equally inclined to another straight line which they cut are parallel.**

Also, the converse of this is clearly true, viz.:

(*b*) **If straight lines are parallel they are equally inclined to another straight line which they intersect.**

Transversal. *A straight line which cuts other straight lines is called a transversal.*

Corresponding angles. The angles ACB, $A'C'B'$, are **corresponding** angles in the congruent triangles ABC, $A'B'C'$ (see § 48). They are also *corresponding angles* when the parallel straight lines BC, $B'C'$ are cut by the transversal PQ; they represent the equal angles of inclination of the parallel straight lines to the transversal.

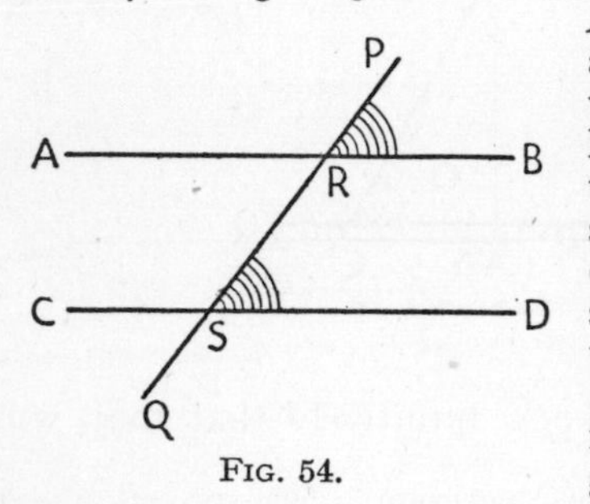

Fig. 54.

53. The conclusions reached above may now be generalised. In Fig. 54, AB and CD are two straight lines cut by a transversal PQ at R and S.

Then $\angle$s PRB, RSD are **corresponding angles.**
$\angle$s PRA, RSC are also corresponding $\angle$s.
From the considerations stated in § 52.

(1) **If AB be parallel to CD.**
Then $\angle PRB = \angle RSD$.

Conversely, (2) **If the corresponding $\angle$s PRB, RSD are equal.**

Then AB and CD are parallel.
These conclusions may be expressed generally as geometrical theorems as follows :

Theorem D. If a straight line cuts two parallel straight lines, corresponding angles are equal.

Theorem D_1 (*converse of previous Theorem*). If two straight lines are cut by another straight line so that corresponding angles are equal, then the two straight lines are parallel.

54. Alternate Angles.

In Fig. 55 the two straight lines *AB*, *CD* are cut by the transversal *PQ* at *R* and *S*. The angles *ARS*, *RSD* are called **alternate angles.**

They lie on alternate sides of *PQ*.

The angles *BRS*, *RSC* also form a pair of alternate angles.

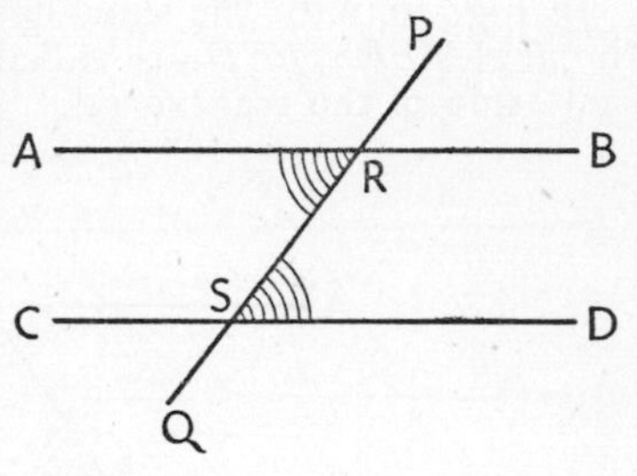

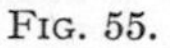
Fig. 55.

(1) Let AB and CD be parallel.

Then, as shown in § 53, corresponding angles are equal.

I.e., $\angle PRB = \angle RSD$,

But $\angle PRB = \angle ARS$ (vertically opp. $\angle$s).

$\therefore \angle ARS = \angle RSD$.

I.e., **The alternate angles are equal.**

Note.—The other pair of alternate angles may similarly be shown to be equal.

Conversely, (2) Let $\angle ARS = \angle RSD$.

Now $\angle ARS = \angle PRB$ (vertically opp. $\angle$s).

$\therefore \angle PRB = \angle RSD$.

But these are corresponding angles.

$\therefore$ **AB is parallel to CD.** (Theor. § 53)

Note.—These results can be expressed in the form of geometric theorems as follows:

Theorem E. If two parallel straight lines are cut by another straight line the alternate angles are equal.

Theorem E_1 (*converse of previous Theorem*). If two straight lines are cut by another straight line so that the alternate angles are equal then the two straight lines are parallel.

55. Interior Angles.

In Fig. 56, with the same figure as in preceding sections the angles BRS, RSD are called the **interior angles on the same side of the transversal.**

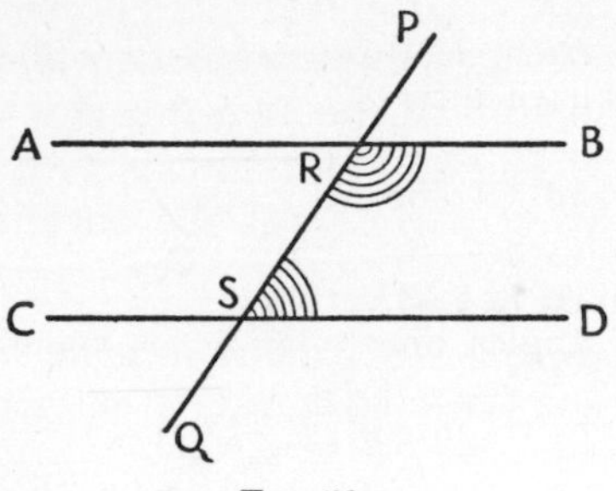

Fig. 56.

Similarly ARS, RSC are interior angles on the other side.

(1) Let AB and CD be parallel.

Then $\angle ARS = \angle RSD$ (alternate angles).
Add $\angle BRS$ to each.
Then $\angle ARS + \angle BRS = \angle RSD + \angle BRS$.
But $\angle ARS + \angle BRS =$ two right angles. (§ 18)
$\therefore$ $\angle BRS + \angle RSD =$ two right angles.

I.e., the sum of the interior angles on the same side of the transversal is equal to two right angles.

Conversely (2). Let

$$\angle BRS + \angle RSD = \text{two right angles.}$$

Then $\angle BRS + \angle RSD = \angle BRS + \angle ARS.$

Subtracting $\angle BRS$ from each side

$$\angle RSD = \angle ARS.$$

But these are alternate angles.

$\therefore$ AB and CD are parallel (§ 53).

These conclusions may be expressed in the form of theorems as follows:

Theorem F. When two parallel straight lines are cut by another straight line the sum of the two interior angles on the same side of the line is two right angles.

Theorem F_1 (*converse of previous Theorem*). When two straight lines are cut by another straight line, and the sum of the two interior angles on the same side of the straight line is two right angles, then the two straight lines are parallel.

56. Summary of above Results.

I. Three properties of parallel straight lines.

If two parallel straight lines are cut by a transversal, then:

D. Pairs of corresponding angles are equal.
E. Pairs of alternate angles are equal.
F. Sum of the interior angles on the same side of the transversal is two right angles.

II. Conditions of parallel straight lines.

Straight lines are parallel if **one** of the following conditions is satisfied:

When they are cut by a transversal, they are parallel:
D_1. *If corresponding angles are equal.*
E_1. *If alternate angles are equal.*
F_1. *If the sum of two interior angles on the same side of the transversal is two right angles.*

57. Construction 2.

To draw through a given point a straight line parallel to a given straight line.

The conditions of E_1 (§ 54) above suggest the method of construction.

Note.—The student is advised to perform the construction which follows, step by step.

Let AB (Fig. 57) be the given straight line and P the given point. It is required to draw through P a straight line parallel to AB.

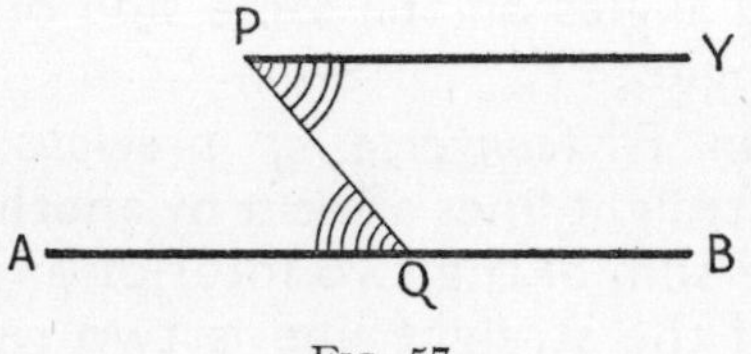

FIG. 57.

Method of construction.

From P draw any straight line PQ to meet AB at Q.

At P make angle QPY equal to $\angle PQA$ (Construction 1, § 25).

Then PY is the straight line required, and it can be produced either way.

Proof. The straight lines PY, AB are cut by a transversal PQ and $\angle YPQ = \angle PQA$. (By construction.)

But these are alternate angles.

$\therefore$ Condition E_1 of § 54 is satisfied.

$\therefore$ PY and AB are parallel.

Notes.—(1) When PY is drawn making $\angle YPQ = \angle PQA$, it is evident that only one such line can be drawn. Hence we conclude:

Through a point only one straight line can be drawn parallel to a given straight line.

(2) If the straight line PQ in Fig. 57 and again in Fig. 58 were drawn perpendicular to AB, since $\angle QPY = \angle PQA$ (§ 54), it is also perpendicular to XY. Hence it may be concluded that:

A straight line which is perpendicular to one of two parallel straight lines, is also perpendicular to the other.

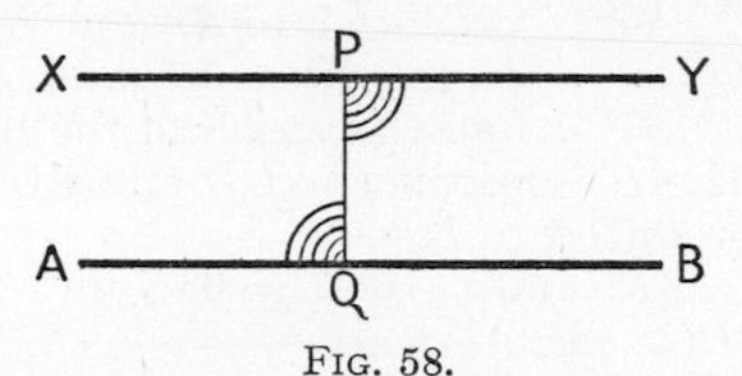

FIG. 58.

Exercise 4

1. In Fig. 59 AB and CD are parallel straight lines and are cut by a transversal PQ at X and Y.

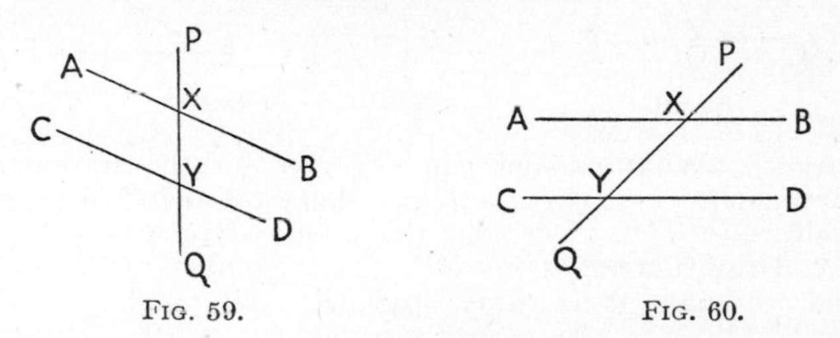

FIG. 59. FIG. 60.

State:

(1) Which are pairs of equal corresponding angles.
(2) Which are pairs of equal alternate angles.
(3) Which are the pairs of interior angles whose sum is two right angles.

2. In Fig. 60 AB and CD are parallel straight lines cut by a transversal at X and Y.

If $\angle PXB = 60°$, find in degrees all the other angles in the figure.

3. In Fig. 59 if the angle $CYP = 60°$ state the number of degrees in the following angles:

(1) PXA, (2) BXY, (3) DYX, (4) QYD, (5) PXB.

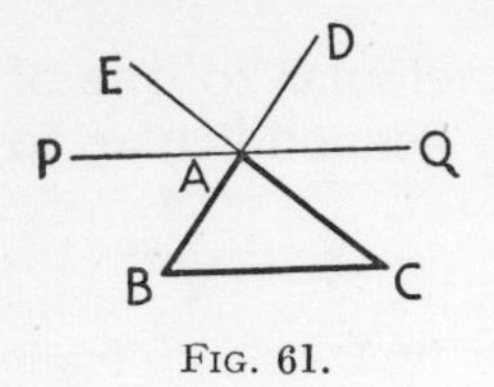

FIG. 61.

4. Through the vertex A of the $\triangle ABC$ (Fig. 61), PQ is drawn parallel to the base BC, and BA and CA are produced to D and E respectively. If $\angle DAQ = 55°$ and $\angle EAP = 40°$, find the angles of the triangle.

5. In Fig 62 AB is parallel to CD and AC is parallel to BD. CD is produced to E.

If $\angle CAD = 37°$ and $\angle BDE = 68°$, find the angles of the figure $ABDC$.

6. In Fig. 63 the arms of the angles ABC, DEF are parallel. What reasons can you give for the statement that $\angle ABC = \angle DEF$? (*Hint.*—Let BC cut DE in Q and produce to R.)

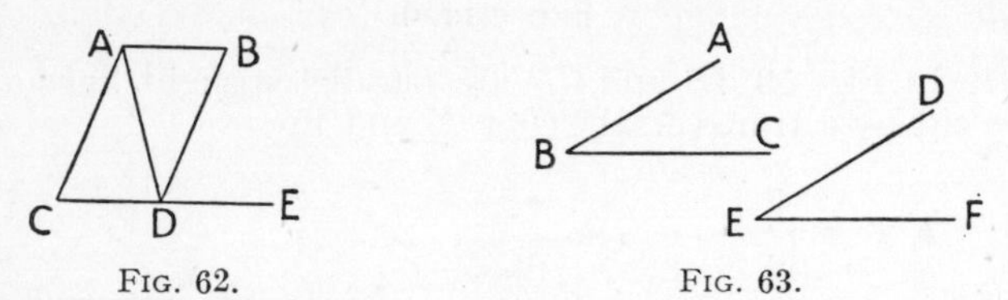

FIG. 62. FIG. 63.

7. Draw a straight line AB. Take points C, D, E on it, and through them draw parallel straight lines making corresponding angles of $30°$ with AB.

8. AB and CD are parallel straight lines (Fig. 64). The angle $APO = 45°$ and $\angle OQC = 35°$. Find the angle POQ.

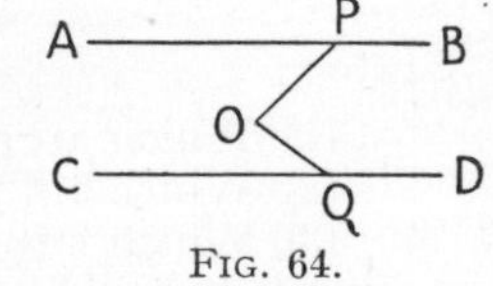

FIG. 64.

9. Two parallel straight lines AB, CD are cut by the transversal PQ at E and F. The $\angle$s BEF and EFC are bisected by the straight lines EH, FG. Prove that these straight lines are parallel.

CHAPTER 8

ANGLES OF A TRIANGLE

58. On several occasions in the previous work the attention of the student has been called to the sum of the three angles of particular triangles. He also has continually before him the triangles represented by the two set squares which he uses and the sum of their angles. It is probable, therefore, that he has come to the conclusion that **the sum of the angles of a triangle is always equal to two right angles or 180°.**

A simple experiment will help to confirm this. Draw any triangle and cut it out. Then tear off the angles and fit them together, as is indicated in Fig. 65.

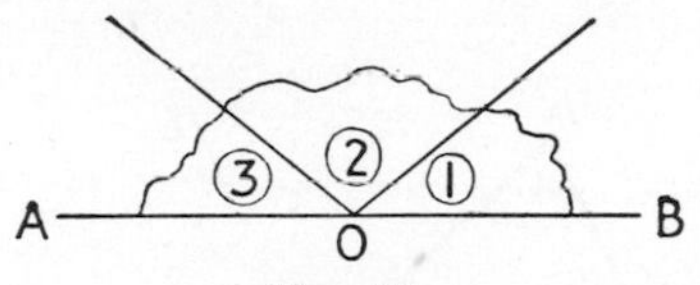

Fig. 65.

The common vertex O will be found to lie in a straight line, AOB. Therefore, as stated in the Theorem of § 18, the sum of the angles—*i.e.*, the angles of the triangle—is 180°.

This is one of the most remarkable facts in elementary geometry, but it would not be satisfactory to accept it as being true for **all** triangles because it has been found to be true in certain cases. We must therefore prove beyond doubt that the result holds for all triangles.

The proof which we shall proceed to give has already been anticipated in Question 4, Fig. 61, of Exercise 4. With a small modification this is substantially the standard proof of the theorem. This is as follows.

59. Theorem. The sum of the angles of any triangle is equal to two right angles.

Fig. 66 represents any triangle ABC. It is required to prove that:

$$\angle ABC + \angle BCA + \angle CAB = \text{two right angles.}$$

To obtain the proof some additional construction is necessary.

Construction. (1) Produce one side, *e.g.*, BC to D.
(2) From C draw CE parallel to BA (*Construction* 2).

Proof.

(1) AB and CE are parallels and AC is a transversal.
∴ Alternate angles ACE, BAC are equal (§ 56 E).

(2) AB and CE are parallel and BC is a transversal.
∴ Corresponding angles ECD, ABC are equal (§ 56 D).

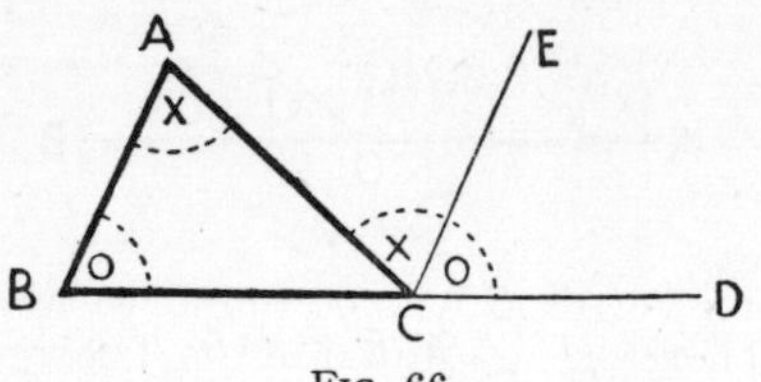

FIG. 66.

(3) ∴ by addition—

$$\angle BAC + \angle ABC = \angle ACE + \angle ECD$$
$$= \angle ACD.$$

(4) Adding $\angle ACB$ to each

$$\angle BAC + \angle ABC + \angle ACB = \angle ACD + \angle ACB$$
$$= \text{two right angles (§ 18).}$$

I.e., the sum of the angles of the triangle is equal to two right angles.

60. It will be evident that in proving the above theorem two other theorems have been incidentally proved. They are as follows:

Theorem. An exterior angle of a triangle is equal to the sum of the two interior opposite angles.

It was shown in step (3) of the proof that

$$\angle ACD = \angle BAC + \angle ABC.$$

I.e., the exterior angle ACD is equal to the sum of the two interior opposite angles ABC and BAC.

Employing a similar proof, it can be shown that if any other side be produced, the exterior angle thus formed is equal to the sum of the two corresponding interior angles.

Theorem. An exterior angle of a triangle is greater than either of the two opposite interior angles.

For, in Fig. 66, since the exterior angle ACD is equal to the sum of the two interior angles ABC, BAC, it must be greater than either of them.

61. Corollaries.

When a theorem has been proved it sometimes happens that other theorems follow from it so naturally that they are self-evident. They are called corollaries.

From the above Theorem the following corollaries thus arise.

(1) *In a right-angled triangle the right angle is the greatest angle.* The sum of the other two must be 90° and each of them is acute.

(2) *No triangle can have two of its angles right angles.*

(3) *In any triangle at least two of the angles must be acute angles.*

(4) *Only one perpendicular can be drawn from a point to a straight line.* This follows from Corollary 3.

Exercise 5

1. Find the third angle of a triangle in which two of the angles are as follows:

(*a*) $87°, 35°$. (*b*) $105°, 22°$.
(*c*) $90°, 46°$. (*d*) $34°, 34°$.

2. In the triangle ABC, the angle A is $40°$ and the angles B and C are equal. Find them.

3. In the $\triangle ABC$, BC is produced to D, and the exterior angle so formed, ACD, is $112°$. If $\angle A = 40°$, what is the angle B?

4. If the three angles of a $\triangle$ are equal, how many degrees are there in each?

5. The angles of a $\triangle$ are in the ratio $1 : 2 : 3$. Find the angles in degrees.

6. In a right-angled triangle two of the angles are equal. Find them in degrees.

7. ABC is a right-angled triangle and $\angle ABC$ is the right angle. From B a perpendicular BD is drawn to AC. If $\angle BAC = 55°$, find the angles ABD, CBD and ACB.

8. In the $\triangle ABC$, from P, a point on AB, PQ is drawn parallel to BC, meeting AC in Q. If the $\angle APQ = 70°$ and $\angle ACB = 50°$, find the angles ABC, AQP, and BAC.

9. In the $\triangle ABC$ the angle B is bisected, and the bisector meets AC in D. If $\angle ABC = 80°$ and $\angle BDC = 95°$, find the angles at A and C.

10. AB and CD are parallel straight lines, and PQ is a transversal cutting them at P and Q. The interior angles at P and Q on the same side of PQ are bisected by the straight lines PR, QR. Show that the angle at R is a right angle.

11. The side BC of the triangle ABC is produced both ways to D and E. If $\angle ABD = 124°$ and $\angle ACE = 130°$, find the angles of the triangle.

12. ABC is a right-angled triangle with B the right angle. BC and CA are produced on to D and E respectively. Prove that $\angle ACD + \angle BAE =$ three right angles.

CHAPTER 9

ISOSCELES TRIANGLES

62. Relations between the Sides and Angles.

In § 42 an isosceles triangle was defined as **a triangle having two equal sides.**

Take two set squares (30°, 60°, 90°) of the same size and place them side by side as in Fig. 67. It will be seen that:

(1) Since $\angle$s ADB, ADC are right angles, **BD and DC are in the same straight line.**

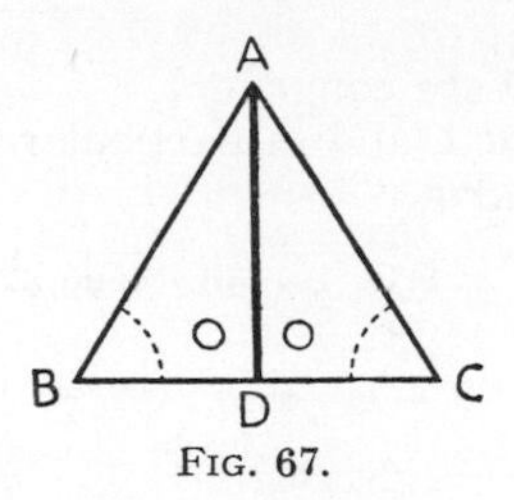

FIG. 67.

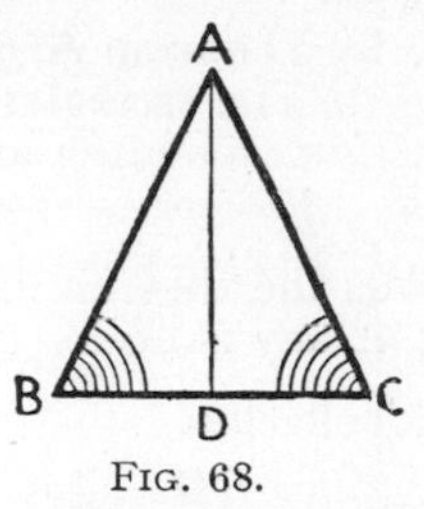

FIG. 68.

(2) The two $\triangle$s together constitute a new $\triangle$ in which $AB = AC$—*i.e.*, **$\triangle$ABC is isosceles.**

(3) $\angle ABC = \angle ACB$, each being 60°, *i.e.*, **the angles which are opposite to the equal sides are equal.**

The question now arises, is this true in all cases? Can we prove the following theorem?

Theorem. If two sides of a triangle are equal, the angles which are opposite to them are equal.

Let ABC be a triangle in which $AB = AC$ (Fig. 68).

We want to prove that

$$\angle ABC = \angle ACB.$$

Fig. 67 suggests the proof might be obtained by drawing AD, **bisecting the angle BAC** and meeting BC in D.

Thus the $\triangle ABC$ is divided into two $\triangle$s, ABD, ACD. If it can be proved that these are congruent by applying one of the three conditions A, B, C of § 49, then it will follow that the $\angle$s ABD, ACD must be equal.

We therefore set down which of the angles and sides of these $\triangle$s are known to be equal.

They are as follows:

In $\triangle$s ABD, ACD.

(1) $AB = AC$ (given).
(2) $\angle BAD = \angle CAD$ (angle BAC was bisected).
(3) AD is a side of each of the $\triangle$s and therefore equal in each triangle, or, as we say, **AD is common to both $\triangle$s.**

$\therefore$ in the two triangles, two sides and the included angle are equal.

$\therefore$ By Theorem A, § 49.

The **triangles ABD, ACD are congruent.**

$\therefore$ Corresponding angles are equal, and in particular

$$\angle ABD = \angle ACD.$$

Thus the theorem is proved to be true for all cases since ABC is any isosceles triangle.

Corollaries.

1. Since $BD = DC$ and $\angle ADB = \angle ADC$.

I.e., these angles are right angles, it can be said that:

In an isosceles triangle the straight line which bisects the angle at the vertex also bisects the base at right angles.

2. *Triangles which are equilateral are also equiangular.*

3. *The perpendicular drawn from the vertex to the base bisects it.*

63. The following theorem follows directly from the Theorem of § 62:

Theorem. If the equal sides of an isosceles triangle are produced the exterior angles so formed are equal.

Let the sides of the isosceles $\triangle ABC$ be produced to D and E (Fig. 69).

Then, the exterior angles *DBC*, *ECB* are supplementary to the angles *ABC*, *ACB* which have been proved equal.

$$\therefore \angle DBC = \angle ECB.$$

64. The converse of the above Theorem is also true. In that Theorem we proved that " if the sides were equal,

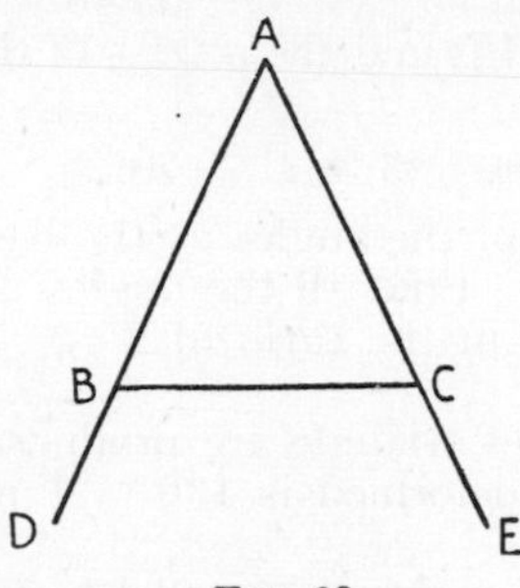

FIG. 69.

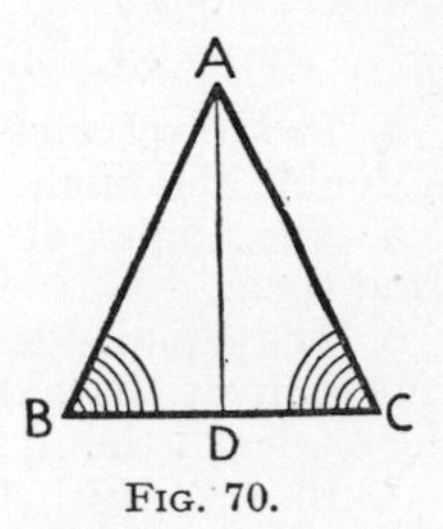

FIG. 70.

the opposite angles must be equal." The converse is, " if the angles are equal, the opposite sides are equal."

Theorem. If two angles of a triangle are equal, the sides opposite to them are equal.

In Fig. 70 the angles *ABC*, *ACB* are given equal.

We require to prove AB = AC, these being the sides opposite to them.

The proof is similar to that of the Theorem above, but Theorem C of § 49 is used instead of Theorem A.

Construction. Draw *AD* bisecting the angle *BAC*.

Proof. In the $\triangle$s ABD, ACD:

(1) $\angle ABD = \angle ACD$ (given).
(2) $\angle BAD = \angle CAD$ (construction).
(3) Side *AD* is common to both.

$\therefore$ $\triangle$s ABD, ACD are congruent (Theorem C, § 49).

In particular AB = AC.

Corollary. *Triangles which are equiangular are also equilateral* [converse of Corollary 2 (§ 62)].

Exercise 6

1. In the isosceles triangles, in which the angle of the vertex is (*a*) 45°, (*b*) 110°, (*c*) 90°, find the remaining angles.

2. Find the angle at the vertex of an isosceles triangle when each of the equal angles is (*a*) 50°, (*b*) 32°, (*c*) 45°.

3. In the triangle ABC, $AB = AC$, find the angles of the triangle when

$$(1)\ \angle B = 48°,\ (2)\ \angle A = 80°,\ (3)\ \angle C = 70°.$$

4. In an isosceles triangle each of the angles at the base is double the angle at the vertex. Find all the angles.

5. The angles of a triangle are in the ratio of 2 : 2 : 5. Find them.

6. The equal sides of an isosceles triangle are produced, and each of the exterior angles so formed is 130°. Find the angles of the $\triangle$.

7. In a $\triangle ABC$, $AB = AC$. PQ is drawn parallel to BC and meets the equal sides in P and Q. Prove that the triangle APQ is isosceles.

8. The equal angles of an isosceles triangle ABC—viz., ACB and ABC—are bisected and the bisectors meet at O. Prove that $\triangle OBC$ is isosceles.

9. ABC is an isosceles triangle, and $AB = AC$. AB is produced to D. If $\angle BAC = 50°$, find the angle CBD.

10. Show that if the mid points of the sides of an equilateral triangle are joined, the resulting triangle is also equilateral. What fraction of the whole triangle is it?

11. ABC is a triangle and D is the mid point of BC. DA is drawn. If $DA = DC$ prove that $\angle BAC$ is a right angle.

CHAPTER 10

SOME FUNDAMENTAL CONSTRUCTIONS

65. Before beginning the study of draughtsmanship, engineering and building students and others must first master a number of fundamental constructions. Some of these will be dealt with in this chapter, others will come later. For these constructions only compass and ruler should be employed for the present.

These constructions are placed before the student not only for their practical value, but also because, with the aid of those geometrical theorems which have been studied in previous chapters, it will be possible **to prove that the method of construction is a correct one,** and must produce the desired result. They will thus furnish exercises in geometrical reasoning of which the reader has already had a number of examples. Two examples of constructions have already been introduced (§§ 25 and 57), and the methods of constructing triangles from fixed data were explained in § 45.

66. Construction No. 3.

(*a*) To construct an equilateral triangle on a given base.

(*b*) At a point on a straight line to construct an angle of 60°.

(*a*) *AB* is a straight line on which it is required to construct an equilateral triangle (Fig. 71).

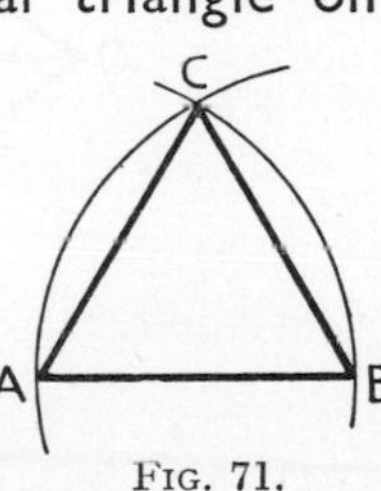

Fig. 71.

Method of construction.

(1) With *A* as centre and *AB* as radius, construct an arc of a circle.

(2) With *B* as centre and *AB* as radius, construct

an arc of a circle large enough to cut the arc previously described in *C*.

(3) Join *AC*, *BC*.

ABC is the required triangle.

Proof. By the method of construction *AC* and *BC* are each equal to *AB*.

∴ they are equal to one another.

i.e. the three sides *AB*, *BC* and *AC* are all equal.

∴ **the triangle ABC is equilateral.**

(*b*) The triangle *ABC* being equilateral is therefore equiangular (Corollary 2, § 62).

∴ each angle is $\frac{1}{3}$ of 180°—*i.e.*, 60°.

∴ **at the points A and B angles of 60° have been constructed.**

67. Construction No. 4.

To bisect a given angle.

Let *AOB* (Fig. 72) be the angle which we require to bisect.

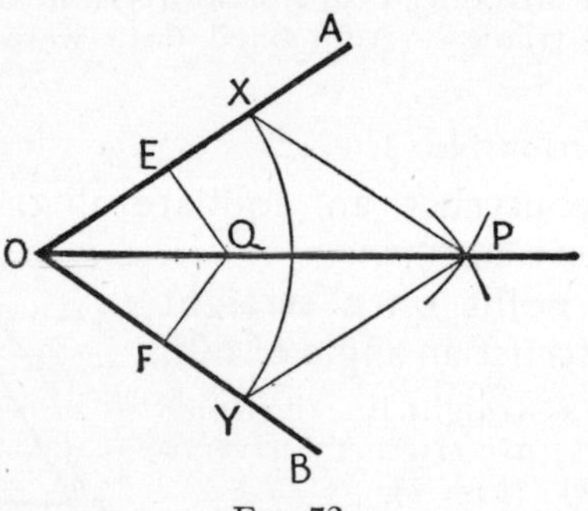

FIG. 72.

Method of construction.

(1) From the two arms of the angle *OA*, *OB* cut off *OX* and *OY* equal to one another.

(2) With X as centre, and with any convenient radius, describe an arc of a circle.

(3) With Y as centre, and the same radius, describe an arc of a circle cutting the other arc in P.

(4) Join OP.

Then OP bisects the angle AOB.

Proof.

Join PX, PY.
In $\triangle$s OPX, OPY:

(1) $OX = OY$ (construction).
(2) $PX = PY$ (construction).
(3) OP is common to both $\triangle$s.

$\therefore$ $\triangle$s OPX, OPY are congruent (§ 49, B).
In particular $\angle POX = \angle POY$.
I.e., OP bisects the angle AOB.

68. This construction suggests the following theorem:

Theorem. Any point on the bisector of an angle is equidistant, from the arms of the angle.

Let Q be any point on the bisector OP of the angle AOB (Fig. 72).
Draw QE and QF perpendicular to the arms OA, OB.
Then QE, QF are the distances of Q from OA and OB.

Proof. In $\triangle$s OEQ, OFQ:

(1) $\angle EOQ = \angle FOQ$ (halves of $\angle AOB$).
(2) $\angle QEO = \angle QFO$ (right $\angle$s).
(3) OQ is common to both $\triangle$s.

$\therefore$ the triangles OEQ, OFQ are congruent (§ 49, C).
In particular $QE = QF$.
$\therefore$ Q is equidistant from the two arms.
Similarly any other point on OP can be shown to be equidistant from OA and OB.

Note.—Students may have noticed that use was made of the bisector of an angle in Theorem of § 62, before the method of obtaining it had been considered. But this does not in any way invalidate the proof of the Theorem above-mentioned since **the bisector of the angle does exist** even though we had not previously proved how it was to be drawn. The proof of the theorem does not in any way depend on the method of drawing an accurate bisector.

69. Construction No. 5.

To bisect a given straight line.

Let AB (Fig. 73) be the straight line which it is required to bisect.

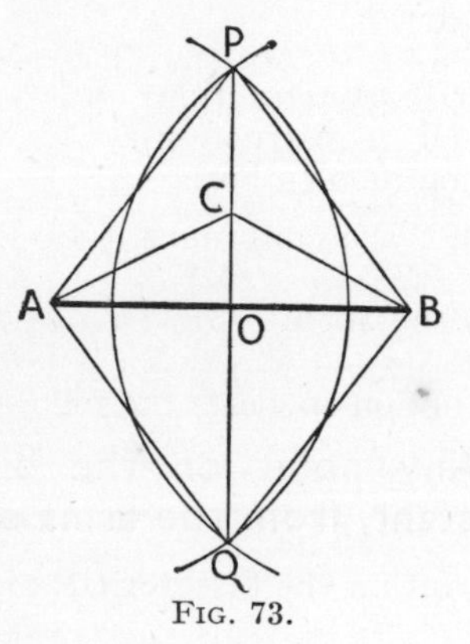

FIG. 73.

Method of construction.

(1) With centres A and B in turn, and a radius greater than $\frac{1}{2}AB$, draw arcs of circles interescting at P and Q.

(2) Join PQ cutting AB at O.

Then **O is the mid point of AB.**

Proof. Join AP, BP, AQ, BQ.

In $\triangle$s APQ, BPQ :

(1) $AP = PB$ (construction).
(2) $AQ = QB$ (construction).
(3) PQ is common to both $\triangle$s.

$\therefore$ $\triangle$s APQ, BPQ are congruent (§ 49, B).

In particular $\angle APQ = \angle BPQ$.

Now the $\triangle APB$ is isosceles and OP bisects the angle at the vertex.

$\therefore$ using the proof of the Theorem of § 62 (Cor. 1).

OP bisects the base AB at right angles.

$\therefore$ AB is bisected at O.

Since OP bisects AB at right angles, this is also the method of the following construction.

To draw the perpendicular bisector of any straight line.

70. A theorem also arises from the above which is similar to that following Construction No. 4, viz.:

Theorem. Any point on the perpendicular bisector of a straight line is equidistant from the ends of the line.

In Fig. 73 if any point C be taken on OP and joined to A and B the $\triangle$s AOC, BOC can be shown to be congruent, as in § 69 and consequently CA = CB.

71. Construction No. 6.

To draw a straight line perpendicular to a given straight line from any point on it.

Let AB (Fig. 74) be a straight line and O a point on it,

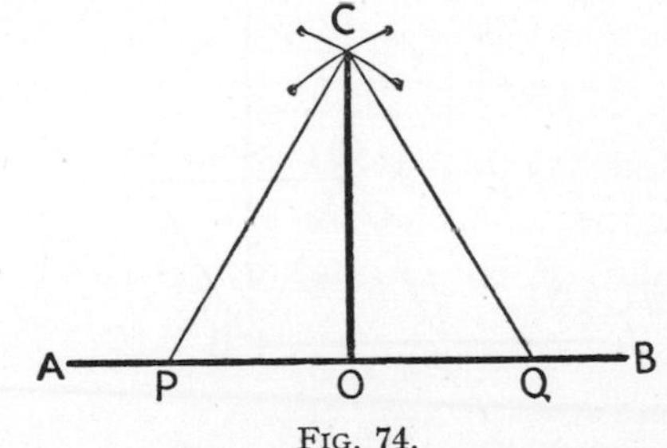

Fig. 74.

at which it is required to draw a straight line which is perpendicular to AB.

Method of construction.

(1) On either side of O mark off equal distances OP, OQ.

(2) With P and Q as centres and any suitable radius describe circles which cut at C.

(3) Join OC.

Then OC is a straight line perpendicular to AB.

Proof. Join CP, CQ.

In $\triangle$s COP, COQ:

(1) $CP = CQ$ (equal radii).
(2) $OP = OQ$ (construction).
(3) OC is common.

$\therefore$ $\triangle$s COP, COQ are congruent (§ 49, B).
In particular $\angle COP = \angle COQ$.
$\therefore$ by definition they are right angles.
$\therefore$ **OC is perpendicular to PQ.**

Note.—If the point O is near one end of AB, so that the two circles cannot conveniently be described, the method of construction No. 7, which follows, can be employed.

72. Construction No. 7.

To draw a perpendicular to a straight line from a point at, or near, one end of it.

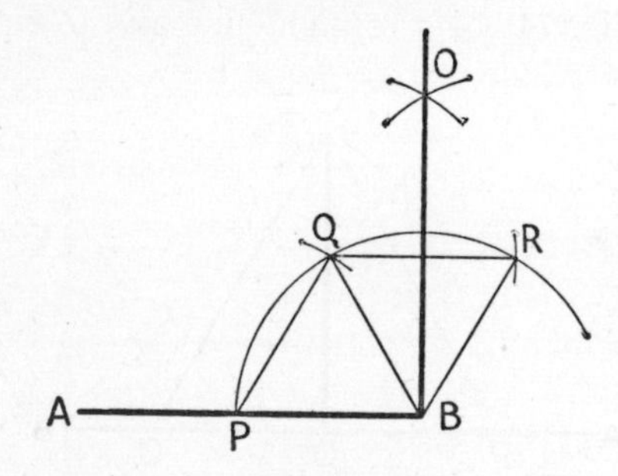

FIG. 75.

Let AB (Fig. 75) be a straight line to which it is required to draw a perpendicular at one end, viz., at B.

Construction.

(1) With centre B and any suitable radius, describe a circle cutting AB at P.

(2) With P as centre and the same radius, describe an arc of a circle cutting the previous circle at Q.

(3) With Q as centre and the same radius, describe an arc cutting the same circle at R.

(4) Join PQ, BQ and BR.

(5) Bisect the angle QBR by OP (Construction No. 4).

Then **OB is perpendicular to AB.**

Proof. Join QR.

As in Construction No. 3, $\triangle BPQ$ is equilateral.

$$\therefore \quad \angle PBQ = 60°.$$

Also BQ, BR, QR are equal.

$\therefore$ $\triangle BQR$ is equilateral and

$$\angle QBR = 60°.$$

Since this is bisected by OB (construction).

$$\therefore \quad \angle OBQ = 30°$$
$$\therefore \quad \angle OBA = 60° + 30° = 90°.$$

$\therefore$ **OB is perpendicular to AB at B.**

73. Construction No. 8.

To draw a straight line perpendicular to a given straight line from a given point without it.

AB (Fig. 76) is a straight line, and P is a point without it. It is required to draw from P a straight line perpendicular to AB.

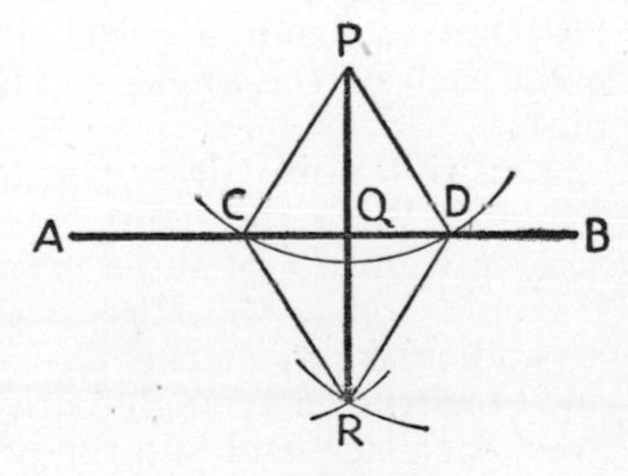

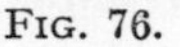
FIG. 76.

Method of Construction.

(1) With P as centre and a convenient radius draw a circle cutting AB at C and D.

(2) With centres C and D and the same radius, a convenient one, draw circles intersecting at R.

(3) Join PR.

Then, **PR is perpendicular to AB.**

Proof. Join PC, PD, RC, RD.

$\triangle$s PCD, RCD are isosceles $\triangle$s on opposite sides of CD.

We can therefore prove as in the proof of Construction No. 5 (Fig. 73), that the $\triangle$s PCQ, PDQ are congruent, and in particular $\angle PQC = \angle PQD$.

$\therefore$ these are right angles and PQ is perpendicular to AB.

Exercise 7

Note.—In the following exercises only a ruler and compasses should be used.

1. Construct the following angles: 30°, 75°, 120°, 150°.

2. Construct on angle of 45°. Use it to obtain an angle of $22\frac{1}{2}$°.

3. Construct the following angles: 15°, 135°, 105°.

4. Construct an equilateral triangle of side 2·5 in. Bisect each side and produce the bisectors. They should meet in a point.

5. Draw a triangle with sides 2 in., 1·5 in. and 1·8 in. Bisect each of the angles. The bisectors should meet in a point.

From this point draw perpendiculars to the sides of the triangle. With the point as centre, and a radius equal to the length of one of these perpendiculars, describe a circle. It should touch the three sides at the points where the perpendiculars meet them.

6. Construct a triangle with sides of 1·9 in., 2·2 in., 2 in. Bisect each side and join the points to the opposite vertices. The three straight lines should be concurrent—*i.e.*, meet in a point.

7. Construct a triangle with sides 2 in., 3 in., 3·5 in. Draw the perpendicular bisectors of the sides. These should meet in a point. With this point as centre, and radius equal to its distance from a vertex, describe a circle. It should pass through the three vertices.

8. Draw a circle of radius 1·8 in. Draw any chord AB and then draw its perpendicular bisector. Repeat this with another chord CD. Do the two perpendiculars meet at the centre of the circle?

9. Draw a triangle of sides 8, 9 and 10 cms. From each vertex draw a perpendicular to the opposite side. The three perpendiculars should be concurrent.

10. AB is a straight line of length 2 in. Show how to find two points each of which is 2·5 in. from both A and B.

CHAPTER 11

QUADRILATERALS

74. As defined in § 38, **a quadrilateral is a plane rectilineal figure bounded by four** straight lines. There are thus four angular points, as A, B, C, D in Fig. 77.

Straight lines which join two opposite angular points are called diagonals.

Thus in Fig. 77, BD is a diagonal, and as A and C can also be joined, every quadrilateral has two diagonals.

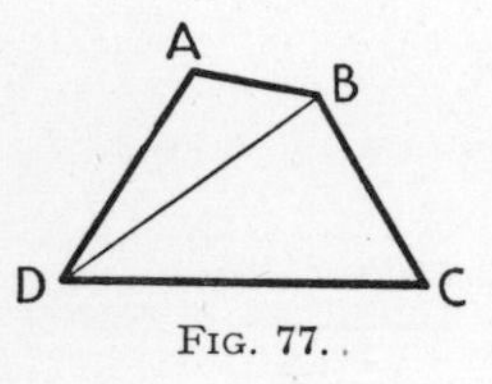

FIG. 77.

Each diagonal divides the quadrilateral into two triangles. Consequently, it follows from Theorem of § 59 that *the sum of the angles of any quadrilateral is equal to four right angles.*

The quadrilateral of Fig. 77 is irregular in shape, but most of the quadrilaterals which we shall consider are **regular quadrilaterals.**

75. Rectangles.

The most commonly occurring quadrilateral in our daily life is the **rectangle.** A knowledge of this figure and its name were assumed in Chapter 1 as being part of the fundamental geometrical knowledge which everybody possesses. In § 2 it appeared again in connection with the solid body, as the shape of each of the faces of a box. The definition of it will come later, as it may be regarded, from the geometrical point of view, as a special form of another quadrilateral which we shall consider next.

76. Parallelograms.

The cover of an ordinary match-box, the inner part having been removed, can be used, as follows, to illustrate a parallelogram. In Fig. 78, *ABCD* represents the open rectangular end of the cover of the box, with the longer sides horizontal. Squeeze gently together the top and bottom of the box so that the sides of the end rectangle rotate until they take up a position such as is shown by *A′B′CD* in Fig. 78.

The opposite sides of A′B′CD are parallel, but its angles are not right angles. The lengths of the sides, however, remain the same.

Such a quadrilateral is called a parallelogram. The original rectangle *ABCD* also has its sides parallel and is

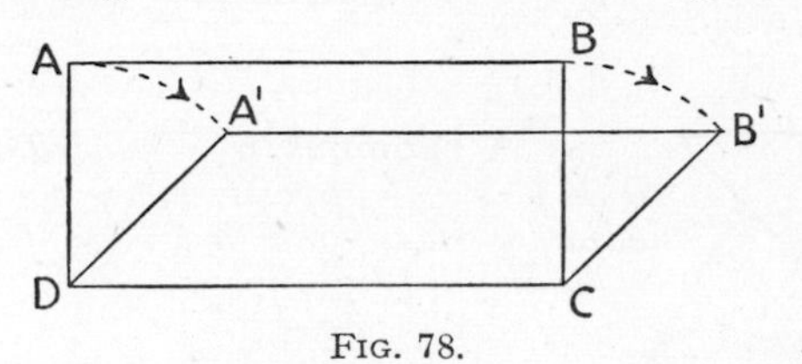

FIG. 78.

therefore, as will be seen later from the definition, a special form of a parallelogram.

If the end of the box had been a square instead of a rectangle, on rotating, it would still be changed to a parallelogram. But just as with the square its sides are all equal. It is therefore a special form of a parallelogram, the **rhombus.**

Considering the four figures, **rectangle, parallelogram, square, rhombus,** it will be seen that they have one property in common, viz., **the pairs of opposite sides are parallel**; they differ however in other respects. These may be contrasted and defined as follows:

77. Definitions of Parallelogram, Rectangle, Square, Rhombus.

(*a*) **A parallelogram** *is a quadrilateral in which both pairs of opposite sides are parallel* (Fig. *a*).

(*a*)

(*a*) Parallelogram.

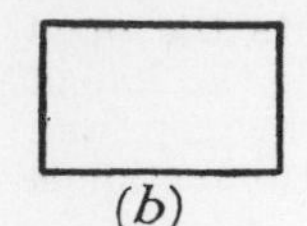

(*b*) Rectangle.

(*b*) **A rectangle** *is a quadrilateral in which both pairs of opposite sides are parallel and one of its angles is a right angle* (Fig. *b*).

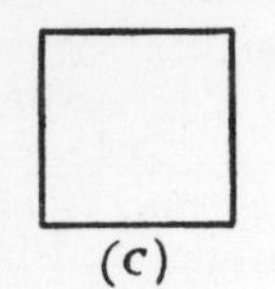

(*c*) Square.

(*c*) **A square** *is a quadrilateral with both pairs of opposite sides parallel, one of its angles a right angle, and two adjacent sides equal* (Fig. *c*).

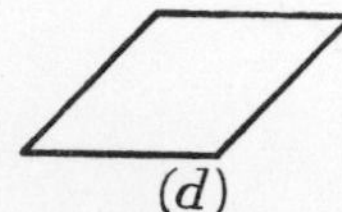

(*d*) Rhombus.
FIG. 79.

(*d*) **A rhombus** *is a quadrilateral with both pairs of opposite sides parallel, two adjacent sides equal but none of its angles right angles* (Fig. *d*).

Notes on the definitions.

(1) The definition of a parallelogram stated above should be examined in connection with the characteristics of a satisfactory definition as described in § 4. Since the opposite sides of a parallelogram are obviously equal there might be a temptation to define it as a " quadrilateral whose opposite sides are equal and parallel ". But the inclusion of the statement of equal sides is illogical. A parallelogram can be constructed by drawing two parallel straight lines and then two other parallel straight lines which cut them. But this construction involves only one geometrical fact about the straight lines, viz., that **they are parallel**. That is all that we know about them. But from this fact, and the properties of parallels which have been considered in Chapter 7, we can proceed to prove that the opposite sides must be equal. This is done in the Theorem of § 78.

(2) It may further be noted that since the rectangle, square and rhombus are all parallelograms, in that they

conform to the definition of having "pairs of opposite sides parallel", they might be defined as follows:

A **rectangle** is a parallelogram in which one of the angles is a right angle.

A **square** is a rectangle which has two adjacent sides equal.

A **rhombus** is a parallelogram with two adjacent sides equal, but none of its angles is a right angle.

(3) In the definition of a rectangle it is stated that "**one** of the angles is a right angle". It will be proved later that if this is so **all the angles must be right angles.** But, for reasons previously given, this does not form part of the definition proper.

78. Properties of Parallelograms.

We can now proceed to establish some of the characteristic properties of the important quadrilaterals dealt with above, basing the proofs upon the definitions given in § 77.

Theorem.

(*a*) The opposite sides and angles of a parallelogram are equal.

(*b*) A parallelogram is bisected by each diagonal.

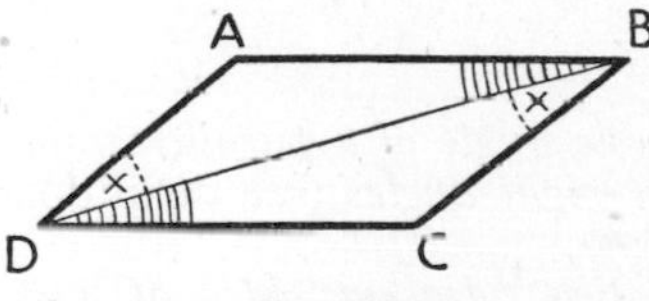

FIG 80.

ABCD (Fig. 80) is a parallelogram and *BD* is one of its diagonals.

(*a*) *We require to prove* $\begin{cases} (1)\ AB = DC. \\ (2)\ AD = BC. \end{cases}$

$\begin{cases} (3)\ \angle ABC = \angle ADC. \\ (4)\ \angle BAD = \angle BCD. \end{cases}$

Proof. By definition AB is parallel to DC
and AD is parallel to BC.

The diagonal BD is a transversal meeting these parallel straight lines.

$\therefore$ In the $\triangle$s ABD, CBD:

(1) $\angle ABD = \angle BDC$ (alternate $\angle$s, § 56).
(2) $\angle ADB = \angle DBC$ (alternate $\angle$s, § 56).
(3) BD is common to both $\triangle$s.

$\therefore$ $\triangle$s ABD, CBD are congruent (§ 49, C).

In particular AB = DC
AD = BC.

Also from (1) and (2) by addition

$$\angle ABD + \angle DBC = \angle ADB + \angle BDC,$$

i.e.,

$$\angle ABC = \angle ADC.$$

Similarly, it may be shown by drawing the other diagonal

that $$\angle DAB = \angle BCD.$$

(*b*) Since the $\triangle$s ABD, CBD are congruent, each of them must be half of the area of the parallelogram, *i.e.*, the diagonal BD bisects the parallelogram.

Similarly, it may be shown that the diagonal AC if drawn would bisect the parallelogram.

Corollaries.

Cor. 1. *If one angle of a parallelogram is a right angle, all the angles are right angles* (see definition of a rectangle, § 77).

Cor. 2. *If two adjacent sides of a parallelogram are equal, all the sides are equal* (see definitions of square and rhombus, § 77).

79. The diagonals of parallelograms.

Theorem. The diagonals of a parallelogram bisect each other.

In Fig. 81, *ABCD* is a parallelogram; *AC* and *BD* its diagonals intersect at *O*.

We require to prove that the diagonals are bisected at *O*, *i.e.*,

$$AO = OC, BO = OD.$$

Proof. In $\triangle$s AOB, COD :

(1) $AB = CD$ (§ 78).
(2) $\angle OAB = \angle OCD$ (alternate $\angle$s, § 56).
(3) $\angle OBA = \angle ODC$ (alternate $\angle$s, § 56).

$\therefore$ $\triangle$s AOB, COD are congruent (§ 49, C).

In particular AO = OC
BO = OD.

Note.—This theorem holds for a rectangle, square and rhombus,

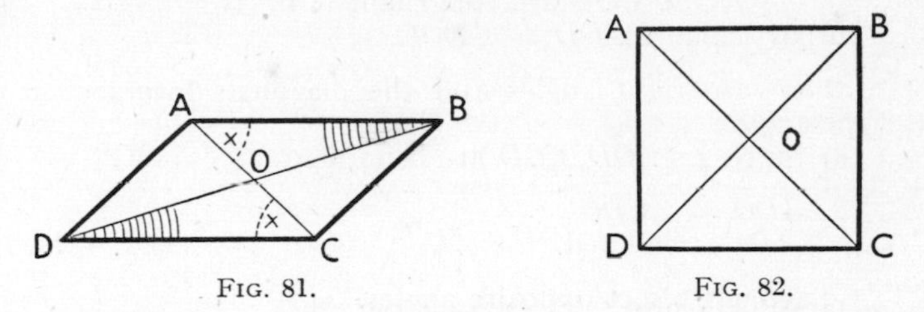

Fig. 81. Fig. 82.

since these are parallelograms and the theorem can be proved precisely as in the above.

80. The diagonals of a square.

Theorem. The diagonals of a square are equal, intersect at right angles and bisect the opposite angles.

ABCD in Fig. 82 is a square and *O* is the intersection of its diagonals.

We require to prove :

(1) The diagonals are equal.
(2) The angles at *O* are right $\angle$s.
(3) The diagonals bisect opposite angles.

Proof.

(1) In the △s ADC, BCD :

(1) $AD = BC$ (§78).
(2) DC is common to each △.
(3) $\angle ADC = \angle BCD$ (right angles, § 78, Cor. 1).

∴ the △s ADC, BCD are congruent (§ 49, A).

In particular $AC = BD$,

i.e., the diagonals are equal.

(2) In the △s AOD, COD :

(1) $AO = OC$ (§ 79).
(2) $AD = DC$ (sides of a square).
(3) OD is common.

∴ △s AOD, COD are congruent (§ 49, B).

In particular $\angle AOD = \angle DOC$.

∴ these are right angles and **the diagonals intersect at right angles.**

(3) Since △s AOD, COD are congruent.

∴ $\angle ADO = \angle CDO$,

i.e., $\angle ADC$ is bisected.

∴ **diagonals bisect opposite angles.**

81. Properties of the diagonals of parallelograms.

The facts deduced above respecting the diagonals of different types of parallelograms may be summarised as follows :

Parallelograms.	Bisect each other.
Rectangles.	(1) Bisect each other.
	(2) Are equal.
Square.	(1) Bisect each other.
	(2) Are equal.
	(3) Are at right angles.
	(4) Bisect opposite angles.
Rhombus.	(1) Bisect each other.
	(2) Are at right angles.
	(3) Bisect opposite angles.

82. The Trapezium.

The trapezium is a quadrilateral in which two opposite sides are parallel, but the other sides are not parallel.

In the quadrilateral *ABCD* (Fig. 83) *AB* is parallel to *DC* but *AD* and *BC* are not parallel.

ABCD is a **trapezium.**

83. The following is a test by which, when the conditions

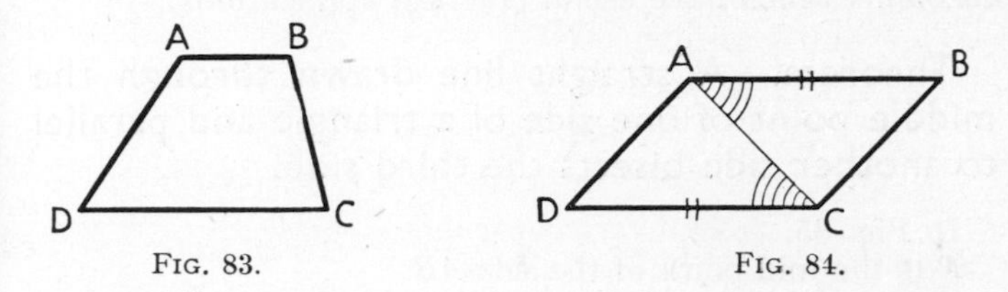

FIG. 83. FIG. 84.

stated are satisfied, a quadrilateral can be declared to be a parallelogram.

Theorem. A quadrilateral, in which one pair of opposite sides are equal and parallel, is a parallelogram.

ABCD (Fig. 84) is a quadrilateral in which *AB* and *CD* are equal and parallel.

Then **ABCD is a parallelogram.**

In order to satisfy the definition of a parallelogram it is necessary **to prove that AD and BC are parallel.**

Construction. Draw the diagonal *AC*.

Proof. **In △s ABC, ADC :**

(1) $AB = CD$ (given).
(2) AC is common.
(3) $\angle BAC = \angle ACD$ (alternate ∠s).

∴ **△s ABC, ADC are congruent** (§ 49, A).

In particular $\angle ACB = \angle DAC$.

But these are alternate angles when the straight lines *AD* and *BC* are cut by the transversal *AC*.

∴ **AD is parallel to BC.** (§ 56, E_1.)

Since AB is parallel to DC.
∴ by the definition of a parallelogram.
ABCD is a parallelogram.

Note.—This theorem may also be stated thus : **The straight lines which join the ends of two equal and parallel straight lines are themselves equal and parallel.**

84. The next two theorems are helpful in proving other theorems which have useful practical applications.

Theorem. A straight line drawn through the middle point of one side of a triangle and parallel to another side bisects the third side.

In Fig. 85.
P is the mid point of the side AB.
PQ is parallel to BC.
We require to prove:
Q is the mid point of AC,
i.e., is $AQ = QC$.

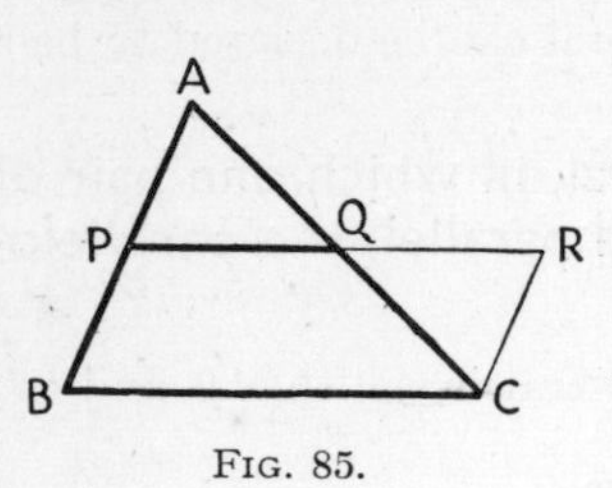

Fig. 85.

The following **construction** is necessary to obtain a proof.

From C draw CR parallel to AB to meet PQ produced in R.

Proof. The opposite sides of the quadrilateral $PRCB$ are parallel.

∴ $PRCB$ is a parallelogram (*Def.*).

∴ $RC = PB$ (§ 78)
$= AP$ (given).

In △s APQ, CRQ :

(1) $AP = RC$ (proved above).
(2) $\angle APQ = \angle QRC$ (alternate ∠s).
(3) $\angle PAQ = \angle QCR$ (alternate ∠s).

∴ **△s are congruent** (§ 49, C).
In particular **AQ = QC.**
∴ **AC is bisected at Q.**

85. Theorem. The straight line joining the middle points of two sides of a triangle is parallel to the third side and is equal to one half of it.

The Fig. 86 is the same as in the preceding theorem, but in this case we are **given** that:

P and Q are **the mid points** of two sides of the $\triangle ABC$.
We require to prove:

(1) PQ is parallel to BC.
(2) $PQ = \frac{1}{2}BC$.

The construction is the same as in the preceding theorem, *i.e.*, CR is drawn parallel to AB to meet PQ produced at R.

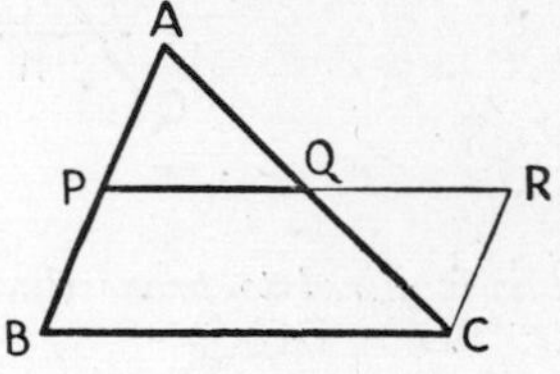

Fig. 86.

Proof. In $\triangle$s APQ, QRC :

(1) $\angle PAQ = \angle QCR$ (alternate $\angle$s).
(2) $\angle APQ = \angle QRC$ (alternate $\angle$s).
(3) $AQ = QC$ (given).

$\therefore$ $\triangle$s APQ, QRC are congruent (§ 49, C).

In particular $AP = RC$
and $PQ = QR$
i.e., $PQ = \frac{1}{2}PR$.
But $AP = PB$ (given).
$\therefore$ $PB = RC$.

But by construction PB and RC are parallel.
$\therefore$ $PRCB$ is a quadrilateral in which a pair of opposite sides is equal and parallel.
$\therefore$ by the Theorem of § 83. PRCB is a parallelogram, *i.e.*, PQ is parallel to BC.
And since $PQ = \frac{1}{2}PR$.
$\therefore$ $PQ = \frac{1}{2}BC$.

86. The following theorem is useful in its practical applications.

Theorem. If three or more parallel straight lines make equal intercepts on any transversal they also make equal intercepts on any other transversal.

In Fig. 87, *AB*, *CD*, *EF* are three parallels. They are cut by two transversals *PQ*, *RS*.

Given that the intercepts on PQ are equal, i.e., $AC = CE$,

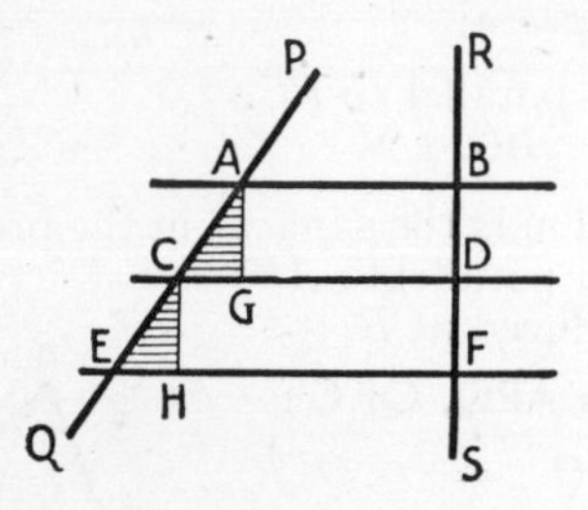

FIG. 87.

it is required to prove that the intercepts on *RS* are equal, *i.e.*, $BD = DF$.

Construction. Draw *AG* parallel to *BD*,
and *CH* parallel to *DF*.

Proof. *AGDB* and *CHFD* are parallelograms (*Def.*).

$\therefore \quad BD = AG$
and $\quad DF = CH$.

In $\triangle$s ACG, CEH :

(1) $AC = CE$ (given).
(2) $\angle ACG = \angle CEH$ (corresponding angles).
(3) $\angle CAG = \angle ECH$ (corresponding angles).

$\therefore$ $\triangle$s ACG, CEH are congruent.

In particular $\quad AG = CH$.
But $\quad AG = BD$,
and $\quad CH = DF$.
$\therefore \quad BD = DF$.

Note.—This theorem is known as "**the theorem of equal intercepts**". It is the basis of the *diagonal scale*.

87. Construction No. 9.

To divide a given straight line into any number of equal parts.

This construction problem is solved by the application of the preceding theorem.

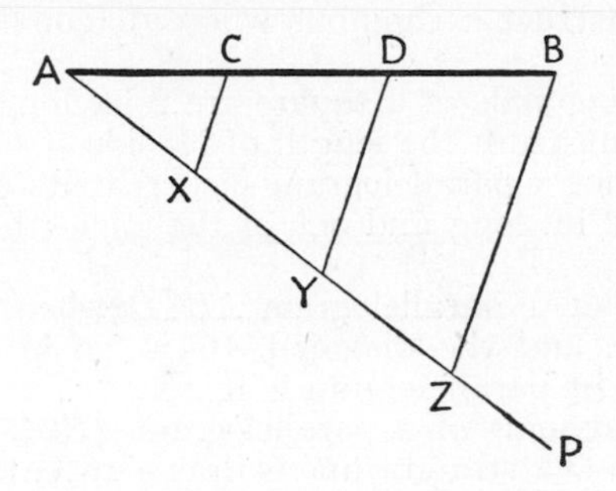

FIG. 88.

In Fig. 88, *AB* is any straight line. Suppose it is desired to divide into (say) **three equal parts,** *i.e.*, **to trisect it.**

Method of construction.

Draw a straight line *AP* making any convenient angle with *AB*.

With a pair of dividers or compasses mark off along *AP* suitable lengths *AX*, *XY*, *YZ*, which are equal.

Join *ZB*.

From *X* and *Y* draw the straight lines *XC*, *YD*, parallel to *ZB*.

The straight line *AB* is trisected at *C* and *D*.

Proof. The two transversals *AB* and *AP* cut the three parallel straight lines *CX*, *DY*, *BZ*.

But the intercepts on *AP*, viz., *AX*, *XY*, *YZ*, are equal.

∴ by Theorem of § 86, the intercepts on AB, viz., AC, CD, DB are equal, *i.e.*, *AB* is trisected at *C* and *D*.

Exercise 8

1. Find the angles of the parallelogram $ABCD$ (Fig. 80) when :

   (*a*) $\angle ADC = 70°$.
   (*b*) $\angle DBC = 42°$ and $\angle BDC = 30°$.

2. Construct a square of 1·5 in. side.

3. (*a*) Construct a rhombus whose diagonals are 1·8 in. and 2·4 in.

   (*b*) The diagonals of a square are 2 in. long. Draw the square and measure the length of its side.

4. Construct a parallelogram such that its diagonals are 1·5 in. and 2 in. long and one of the angles between them is 60°.

5. Construct a parallelogram $ABCD$ when $AB = 2$ in., $BC = 1·5$ in. and the diagonal AC is 2·5 in. What particular form of parallelogram is it?

6. The diagonals of a parallelogram $ABCD$ intersect at O. Through O a straight line is drawn to cut AB and CD at P and Q respectively. Prove that $OP = OQ$.

7. ABC is a $\triangle$. E and F are the mid points of AB and AC. EF is produced to G so that $FG = EF$. Prove that BE is equal and parallel to CG.

8. Draw a straight line 4 in. long and divide it into five equal parts. Check by comparing their lengths by measurement.

9. $ABCD$ is a trapezium in which AB is parallel to CD. If $AD = BC$ prove that $\angle ADC = \angle BCD$.

CHAPTER 12

AREAS OF RECTILINEAL FIGURES

88. **Area** was defined in § 7 as the **amount** of surface enclosed by the boundaries of a figure and there have been several implicit references to the areas of rectilineal figures in preceding chapters. For example, when, in § 78 it was proved that "*a parallelogram is bisected by a diagonal,*" the reference was to area only. Again, when it was stated that "*congruent triangles coincide*" the implied meaning is that not only are corresponding sides and angles equal, but that the areas of such triangles are also equal.

It is now necessary to consider the methods by which the areas of rectilineal figures are obtained and, also, how these areas are measured.

89. Measurement of Area.

The first essential for all measurement is a **unit** and the unit of area must clearly be related to the unit of length. If the **unit of length is an inch,** then the unit of area is that of a square, each of whose equal sides is one inch. Thus, the unit of area is a **square inch.**

If for larger measurements the unit of length is a yard, then the unit of area is that of a square, each of whose sides is a yard, *i.e.*, a **square yard.** For very large measurements, such as the area of a country, the unit is a **square mile.**

In the metric system, when the unit of length is a **centimetre,** the unit of area is a **square centimetre.** For larger areas the square metre or square kilometre would be employed.

90. **Square inches,** true to size, may be seen in what is called "squared paper", such as is used for drawing graphs, etc. An example of one of these square inches is shown in Fig. 89, in which the square marked as $ABCD$ is an exact square inch.

The sides of the square are divided into 10 equal parts,

each part being thus 0·1 in. Each division is the side of a square such as is indicated in the small square at *A*.

Consequently, in the bottom row, corresponding to the side *AB*, there are 10 of these small squares. Throughout the whole square *ABCD* there are 10 such rows, since each side of the square is also divided into 10 equal parts. Altogether then there are 10×10, *i.e.*, 100 small squares such as that at *A*.

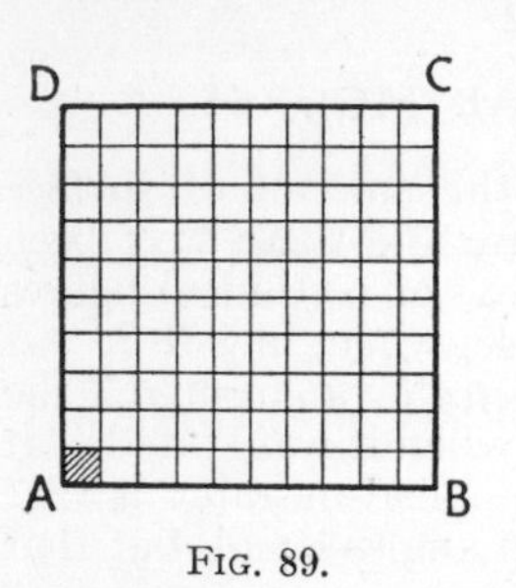

Fig. 89.

Every small square is therefore 0·01 of a square inch. Thus 3 rows contain (10×3) of these and their total area is $0·01 \times 30 =$ 0·3 sq. in.

Similarly 7 rows would contain $(10 \times 7) = 70$ small squares, and the area of the rectangle represented by these 7 rows would be $70 \times 0·01 =$ 0·7 sq. in.

91. Area of a Rectangle.

The above example suggests a method for finding the area of a rectangle. As a more general case let us consider the rectangle shown in Fig. 90, which is drawn on squared paper ruled in centimetres, the square *ABCD* being a square centimetre. Each centimetre is divided into 10 equal divisions, each a millimetre.

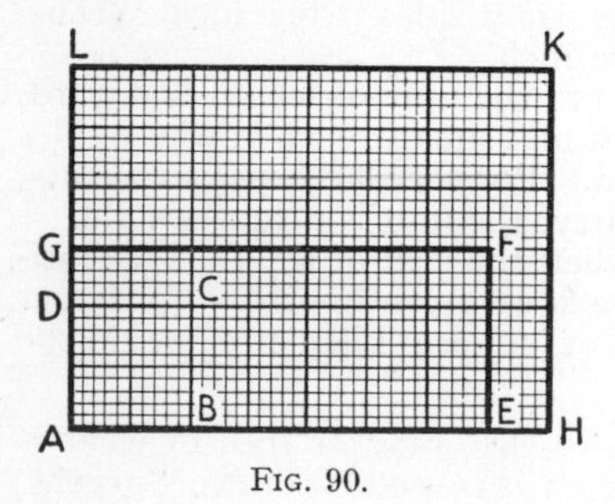

Fig. 90.

The sides of the rectangle *AHKL* are 4 cm. and 3 cm.

Corresponding to each centimetre in the side *AH* there is a sq. cm. above it, *i.e.*, there are 4 sq. cm. in the row of squares constructed above each centimetre along *AH*.

In the whole rectangle *ALKH* there are 3 such rows.

$\therefore$ The total number of sq. cm. in $ALKH = 3 \times 4 = 12$, *i.e.*, the **area of AHKL = 12 sq. cm.**

If the rectangle were 6 cm. by 5 cm., then there would be 6 sq. cm. in each row and 5 rows.

$$\therefore \text{Total area} = 6 \times 5 \\ = 30 \text{ sq. cm.}$$

This reasoning can evidently be applied to a rectangle of any size and the result generalised as follows:

Let a = number of units of length in one side of the rectangle.

Let b = number of units of length in the adjacent side of the rectangle.

Then area of rectangle = (a × b) sq. units.

The argument above referred to examples in which the lengths of the sides of the rectangle are exact numbers of units of length. With suitable modifications, however, it can be shown to be true when the lengths of the sides are fractional.

For example adjacent sides of the rectangle $AEFG$, Fig. 90, viz., AE and AG, are 3·5 and 1·5 cm. respectively.

These lengths expressed in millimetres are 35 and 15 mm. respectively, and each very small square with a side one millimetre is a square millimetre.

$\therefore$ with the same reasoning as above.

$$\text{Area of } AEFG = (35 \times 15) \text{ sq. mm.} \\ = 525 \text{ sq. mm.} \\ = 5.25 \text{ sq. cm.}$$

Similar methods employed in other cases confirm the truth of the general rule given above.

92. Area of a Square.

Regarding a square as a rectangle with adjacent sides equal, the above formula for its area can be modified accordingly.

Thus if $b = a$.

Then **area of square** $= a \times a$

$= a^2$ sq. units.

93. Area of a Parallelogram.

The formula for the area of a parallelogram, the angles of the figure not being right angles, can be determined as follows:

Consider the parallelogram *ABCD*, Fig. 91.

Construction. Draw *CP* perpendicular to *AB*.

From *D* draw *DQ* perpendicular to *CD* to meet *BA* produced at *Q*. *DQ* is perpendicular to *BQ*.

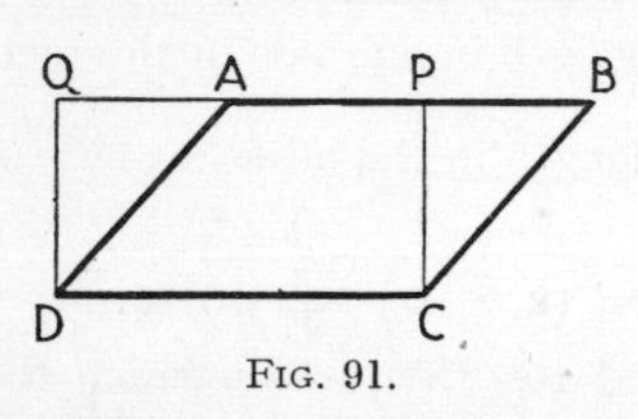

Fig. 91.

Proof. If *CD* be regarded as the base of the parallelogram, then **CP or DQ is the corresponding altitude or height** of the parallelogram. It is the distance between the two parallels *AB* and *DC*.

DCPQ is a rectangle and *DC* and *CP* are two adjacent sides.

In △s BCP, ADQ :

(1) $\angle CBP = \angle DAQ$ (corresponding angles).
(2) $\angle DQA = \angle CPB$ (right angles).
(3) $CB = DA$ (§ 78).

∴ △s BCP, ADQ are congruent.

∴ quadrilateral $ADCP + \triangle PCB =$ quadrilateral $ADCP + \triangle ADQ$,

i.e., parallelogram $ABCD =$ rectangle $PCDQ$.

∴ the area of a parallelogram is equal to the area of the rectangle with the same base and same height.

∴ area of parallelogram = base × height.

Corollary. *The area of any other parallelogram with the base* DC *and having the same height, or lying between the same parallels, is equal to that of the rectangle* PCDQ.

94. The statement in the corollary above can be expressed formally in the following theorem.

Theorem. Parallelograms on the same base and having the same height, or between the same parallels, are equal in area.

In Fig. 92 PQ and XY are two parallels.

Let AB, a part of PQ, be a base to any two parallelograms,

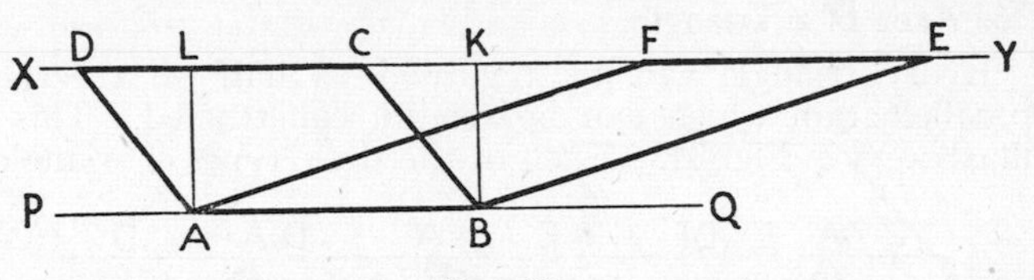

FIG. 92.

such as $ABCD$, $ABEF$, between the two parallels PQ and XY.

Draw AL and BK perpendicular to XY.

Then $$AL = BK.$$

As in § 93 each of the parallelograms $ABCD$, $ABEF$ can be shown to be equal in area to the rectangle $ABKL$.

$\therefore$ **the area of ABCD = area of ABEF.**

95. If both base and height be constant in parallelograms such as are described above, the area will be constant.

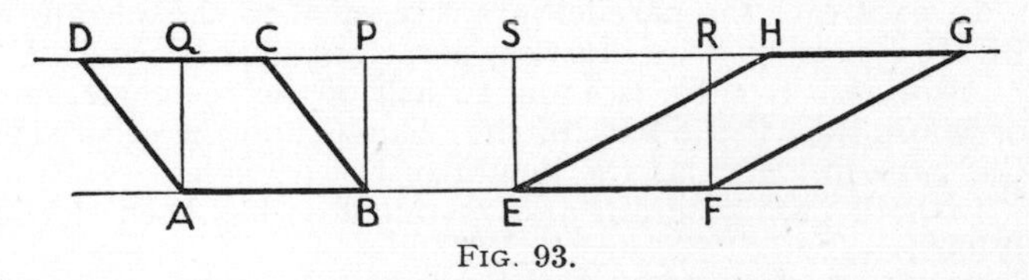

FIG. 93.

The base need not be the **same** base; but the bases must be equal.

Hence we arrive at :

Theorem. Parallelograms which have equal bases and lie between the same parallels, *i.e.*, they have the same height, are equal in area.

In Fig. 93 *ABCD, EFGH* are parallelograms having equal bases *AB* and *EF* and lying between the same parallels.

If rectangles *ABPQ, EFRS* be constructed as shown in figure, these will have the same area.

∴ **parallelograms ABCD, EFGH, which are equal in area to these, must be themselves equal.**

96. Area of a Triangle.

Every triangle can be regarded as half of a certain parallelogram which can be readily constructed. This is illustrated in Fig. 94. Each of the three types of triangles,

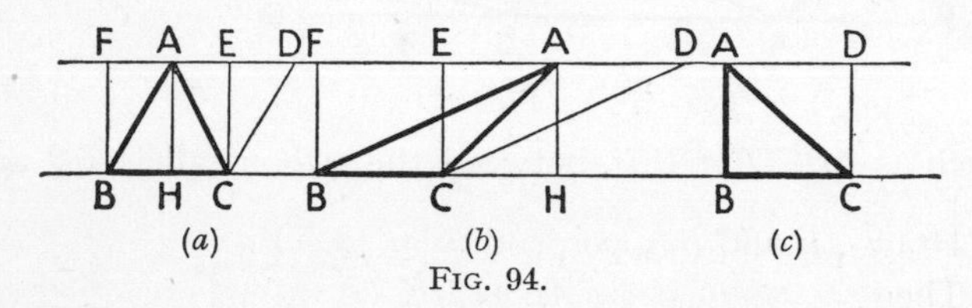

FIG. 94.

acute angled (*a*), obtuse angled (*b*) and right angled (*c*) is half of the parallelogram *ABCD*, the construction of which is obvious. In the case of the right-angled triangle (*c*) the parallelogram assumes the form of a rectangle. In (*a*) and (*b*) *AH* represents the altitude or height of the triangle and therefore also of the corresponding parallelogram.

In **each case the parallelogram is equal to the rectangle BCEF,** constructed by drawing perpendiculars *BF* and *CE*.

Also, **each triangle is equal to half of the rectangle,** one of whose sides is the base of the triangle and the other side the same in length as the height of the triangle.

In both parallelogram and rectangle it has been shown (§ 93) that :

$$\text{Area} = \text{base} \times \text{height}.$$

$$\therefore \textbf{area of } \triangle = \tfrac{1}{2}(\textbf{base} \times \textbf{height}).$$

If b = length of base

h = altitude

and A = area.

Then $\mathbf{A = \tfrac{1}{2}bh}$.

97. From the above conclusions the truth of the following theorems will be apparent without any formal statement of the proofs.

Theorem. If a parallelogram and a triangle be on the same or equal bases and between the same parallels, the area of the triangle is one half that of the parallelogram.

Theorem. Triangles on equal bases and between the same parallels are equal in area.

98. Area of a Trapezium.

ABCD is a trapezium (Fig. 95) in which *AD* is parallel to *BC*.

From *D* and *B* draw perpendiculars *DE* and *BF* to the opposite side, produced in the case of *DA*. Join *BD*.

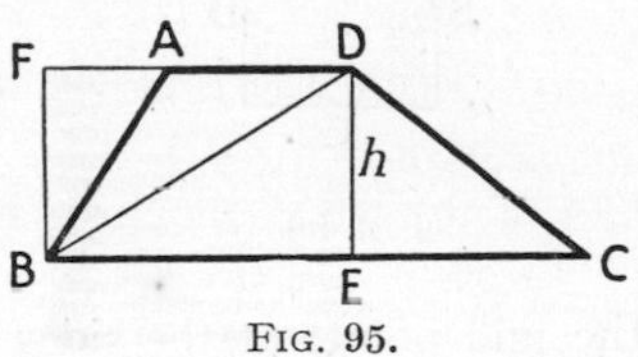

FIG. 95.

The trapezium is divided by *BD* into two $\triangle$s *ABD*, *DBC*.

Let h be the distance between the parallel sides.

Then h is equal to *DE* and *BF*, the altitudes of the $\triangle$s *DBC* and *ADB*.

Let $AD = a$ units of length and $BC = b$.

Area of $\triangle DBC = \frac{1}{2}bh$.

Area of $\triangle ADB = \frac{1}{2}ah$.

$$\therefore \text{ area of trapezium} = \tfrac{1}{2}ah + \tfrac{1}{2}bh$$
$$= \tfrac{1}{2}h(a + b)$$
$$= \tfrac{1}{2}(\text{height} \times \text{sum of parallel sides})$$
or
$$= \text{height} \times \text{average of parallel sides}.$$

Area of a Quadrilateral. Any quadrilateral can be divided into two triangles, as in the trapezium above. The sum of the areas of these triangles is equal to the area of the quadrilateral.

Exercise 9

Note.—In some of the following exercises the student is expected to draw the figures to scale and calculate the areas from the measured lines.

1. Take two set squares of angles 30°, 60°, 90° and place them together with the hypotenuses coinciding, thus forming a rectangle. Measure the sides of the rectangle and find its area. Hence find the area of one of the set squares.

2. Fig. 96 represents a square tile of side 5 in. *ABCD* are

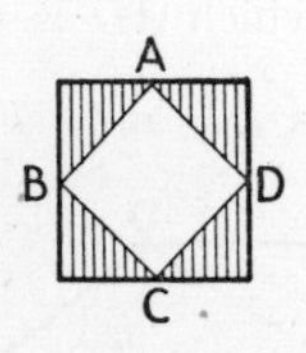

FIG. 96.

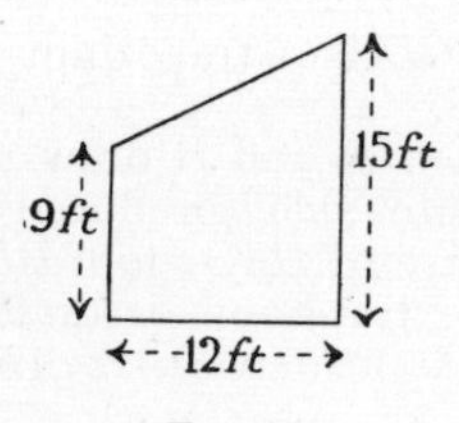

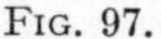

FIG. 97.

the mid points of the sides of the square. Find the area of the part which is shaded.

3. Find the area of a rectangle 5·8 cm. by 4·5 cm.

4. Find the areas of the following triangles :

   (*a*) Base 15 ft. 6 in., height 7 ft.
   (*b*) Base 9·7 in., height 6·7 in.
   (*c*) Base 15·4 cm., height 11·4 cm.

5. Construct a triangle with sides 2, 2·5 and 3 in. and find its area. Check the result by doing it in three ways and finding the average of your results.

6. Construct an equilateral triangle of side 7 cm. and find its area.

7. The diagonals of a rhombus are 3·6 in. and 1·4 in. Find the area of the rhombus.

8. The area of a triangle is 15·6 sq. cm. and the length of a base is 6·5 cm. Find the corresponding altitude.

9. Fig. 97, not drawn to scale, represents the side of a lean-to shed of dimensions as indicated. Find its area.

10. Fig. 98, not drawn to scale, represents the section

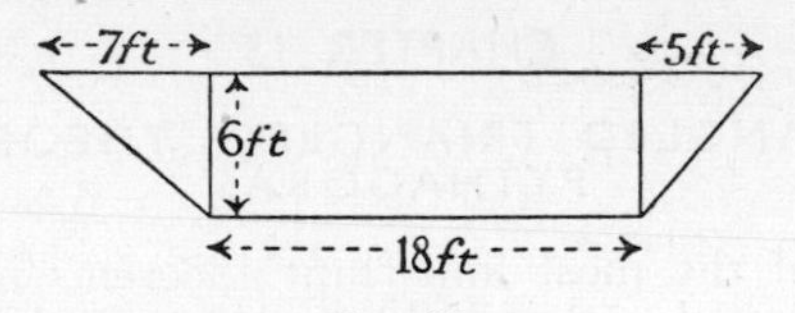

FIG. 98.

of an aqueduct of dimensions as indicated. Find the area of the section.

CHAPTER 13

RIGHT-ANGLED TRIANGLES. THEOREM OF PYTHAGORAS

99. One of the most important theorems in Geometry is that connected with a right-angled triangle and known as the " Theorem of Pythagoras ". It is as follows:

Theorem. The area of the square on the hypotenuse of a right-angled triangle is equal to the sum of the areas of the squares on the other two sides.

In Fig. 99 *ABC* is a right-angled △ and *BC* is its hypotenuse. Squares are constructed on the sides.

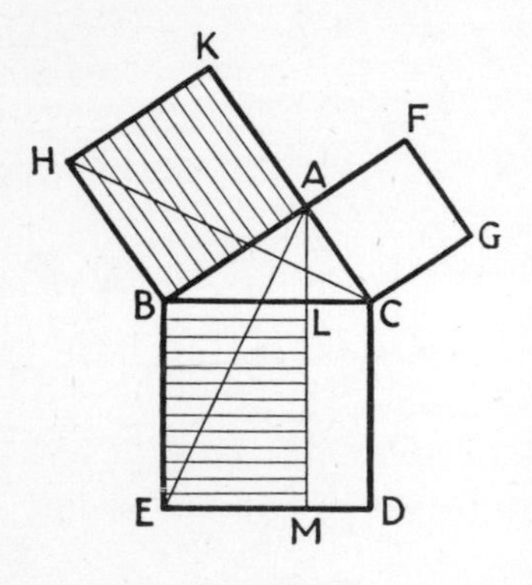

FIG. 99.

The Theorem states that :

Square on BC = square on AB + square on AC.

There are several interesting devices for cutting up the squares on *BA* and *AC* and fitting them into the square on *BC*. But to save time and space we will proceed directly to the proof which the student may find longer and perhaps more complicated than previous theorems.

Construction. Draw *ALM* parallel to *BE* and *CD*, and consequently at right angles to *BC* and *ED*.

This divides the square on *BC* into two rectangles.

The proof consists in showing that

rectangle BLME = square on AB.

(these are shaded) and afterwards

the rectangle CDML = square on AC.

(A) To prove rectangle BLME = square on AB.

To obtain connecting links between these, join AE, HC.

First it is noted that KA and AC are in the same straight line (§ 18).

(1) $\triangle ABE$ and rectangle $BLME$ are on the same base BE and between the same parallels BE and AM.

$\therefore$ $\triangle$ABE = $\frac{1}{2}$(rectangle BLME) (§ 97).

(2) $\triangle HBC$ and square $ABHK$ are on same base HB and between the same parallels HB and KC.

$\therefore$ $\triangle$HBC = $\frac{1}{2}$(square ABHK).

$\therefore$ *If it can be proved that* $\triangle$ABE = $\triangle$HBC.

Then rectangle BLME = *square* ABHK.

(B) To prove $\triangle$ABE = $\triangle$HBC.

(1) $AB = HB$ (sides of a square).
(2) $BE = BC$ (sides of a square).
(3) $\angle ABE = \angle HBC$ (since each $\angle$ equals a right $\angle + \angle ABC$).

$\therefore$ $\triangle$s ABE, HBC are congruent. (§ 49 A).

$\therefore$ rectangle BLME = square ABHK . (1)

In a similar manner by joining AD and BG it may be proved that the

rectangle CLMD = square ACGF . (2)

By addition of (1) and (2) we get

square on BC = sum of squares on AB and AC.

The student is advised to go through the proof leading to (2) and write it down.

The above proof has been made somewhat longer by explanations designed to help the student through it. A shorter and more usual way of setting it out is given in Part II (p. 265).

100. The *converse* of the above theorem is also true. It is as follows:

Theorem. If in a triangle the square on one side is equal to the sum of the squares on the other two sides, then the triangle is right-angled.

The proof is omitted in this section of the book, but will be found in Part II (p. 267).

101. The Theorem of Pythagoras can be expressed in algebraical form as follows:

In the right-angled triangle ABC (Fig. 100) the sides are represented by a, b, c as explained in § 44, c being the hypotenuse.

Then by the Theorem of Pythagoras:

$$c^2 = a^2 + b^2,$$

whence $$c = \sqrt{a^2 + b^2}$$

also $$a^2 = c^2 - b^2$$

and $$a = \sqrt{c^2 - b^2}$$

similarly $$b = \sqrt{c^2 - a^2}.$$

Hence any one side can be expressed in terms of the other two.

From these results it is evident that if in two right-angled triangles, the hypotenuse and one side of each are equal, the third side will also be equal and the triangles are congruent.

102. To find the length of the diagonal of a square in terms of the sides.

Let a = length of a side of the square $ABCD$ (Fig. 101).
Let x = length of the diagonal, DB.

Then $$x^2 = a^2 + a^2 \text{ (Theorem of Pythagoras)}$$
$$= 2a^2.$$
$$\therefore \quad x = a\sqrt{2}.$$

I.e., the ratio of the diagonal of a square to a side is $\sqrt{2} : 1$.

It may be noted that the angles CBD, CDB are each 45°.

103. To find the height, or altitude, of an equilateral triangle in terms of the side.

In Fig. 102, ABC is an equilateral $\triangle$.

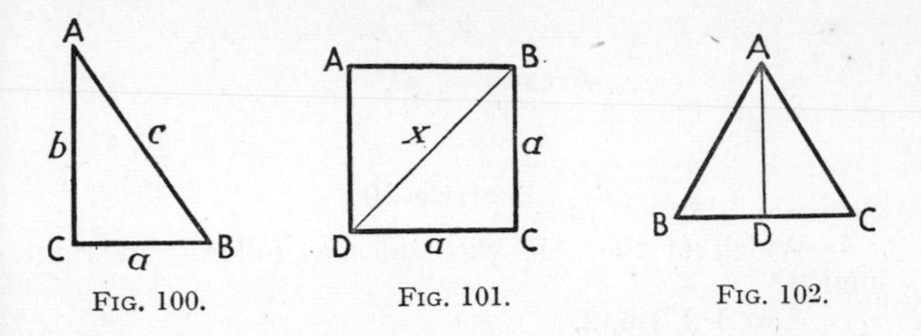

FIG. 100. FIG. 101. FIG. 102.

AD, the perpendicular from A to BC, is the height or altitude.

Let a = length of each side.

Then $CD = \frac{a}{2}$.

In the right-angled $\triangle ADC$, $AC^2 = AD^2 + DC^2$.

$$\therefore \quad AD^2 = AC^2 - DC^2$$
$$= a^2 - \left(\frac{a}{2}\right)^2$$
$$= \frac{3a^2}{4}$$
$$\therefore \quad AD = \frac{\sqrt{3}}{2}a.$$

The angles of the $\triangle ADC$ are 90°, 60°, 30° (those of one of the set squares) and the $\triangle$ is one of frequent occurrence. It should be noted that the ratio of the sides of this $\triangle$ are

$$a : \frac{\sqrt{3}}{2}a : \frac{a}{2}$$

or $$2 : \sqrt{3} : 1.$$

Area of an equilateral triangle.

From the above triangle:

$$\text{Area} = \tfrac{1}{2}BC \times AD$$
$$= \frac{a}{2} \times \frac{\sqrt{3}}{2}a.$$
$$\therefore \quad \text{Area} = \frac{\sqrt{3}}{4}a^2.$$

Exercise 10

1. Which of the Δs, with sides as follows, are right-angled?

   (*a*) 1·2, 1·6, 2.
   (*b*) 4, 5, 6.
   (*c*) 1, 2·4, 2·6.
   (*d*) 5, 7, 9.

2. Find the lengths of the diagonals of squares whose sides are:

   (*a*) 1 in., (*b*) 12 in.

3. Find the altitudes of equilateral Δs whose sides are:

   (*a*) 1 in., (*b*) 12 in.

4. The diagonals of a rhombus are 4·6 in. and 5·2 in. Find the lengths of the sides.

5. A man travels 15 miles due east and then 18 miles due north. How far is he from his starting point in a straight line?

6. One side of a rectangular field is 140 yds. A diagonal is 160 yds. Find the length of the other side.

7. A peg is 15 ft. from the foot of a flagstaff which is 40 ft. high. What length of rope, when taut, will be needed to reach the peg from the top of the flagstaff?

8. The hypotenuse of a right-angled triangle is 6·5 cm. and one side is 2·5 cm. What is the length of the other side?

9. Find the altitude of an isosceles triangle in which each of the equal sides is 10 in., and the base is 5 in.

10. (*a*) Find the area of an equilateral triangle of side 5

in.; (*b*) If the area of an equilateral triangle is $25\sqrt{3}$ sq. in., what is the length of each side?

11. A ladder 40 ft. long rests against a vertical wall of a house, so that the foot of the ladder is 14 ft. from the bottom of the wall. How far up the wall does the top of the ladder reach?

12. The diagonals of a quadrilateral intersect at right angles. Prove that the sum of the squares on one pair of opposite sides equals the sum of the squares on the other pair.

13. Construct a square so that its area is twice that of a given square.

14. A rod 3 ft. in length makes an angle of 30° with its projection on a horizontal plane. Find the length of its projection.

CHAPTER 14

POLYGONS

104. A polygon *is a plane rectilineal figure bounded by more than four straight lines.*

Triangles and quadrilaterals are sometimes included under the term polygon, but it will be used here in the sense defined above.

A regular polygon *is one in which all the sides and all the angles are equal.*

A convex polygon, like a convex quadrilateral, *is one in which no angle is greater than two right angles, i.e.,* it has no reflex or re-entrant angle.

The work which follows will be confined to regular convex polygons.

105. Sides of Polygons.

There is no theoretical limit to the number of sides of a polygon, but only those with twelve or less are commonly met with: The names of polygons which are most in use are as follows:

No. of Sides.	Name.
5	Pentagon
6	Hexagon
7	Heptagon
8	Octagon
10	Decagon

The number of angles of a polygon is the same as the number of sides. Thus a **regular hexagon** has **six** equal angles, as well as six equal sides.

106. Circumscribing Circles of Polygons.

With all regular polygons, circles can be described which pass through all the angular points or vertices.

Such circles are called **circumscribing circles.** Examples are shown in Fig. 103.

To draw the circumscribing circle. If it is required to

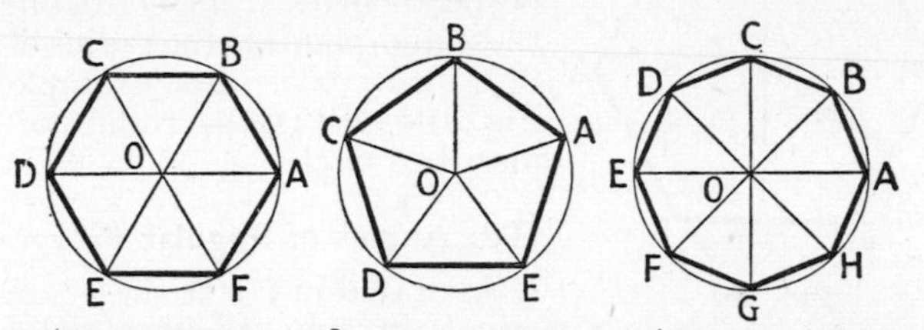

(a) Regular Hexagon (b) Regular Pentagon (c) Regular Octagon

FIG. 103.

draw the circumscribing circle of a given regular polygon, this can be done in two ways.

(1) Draw the perpendicular bisectors of two or more sides; or

(2) Draw the bisectors of two or more of the angles of the polygon.

In either case the intersection of the lines so obtained is the centre of the required circle. In the examples of Fig. 103 all the lines thus drawn are bisectors of the angles. Each of these straight lines is a radius of the circle.

The proof in either case will be clear from previous work.

Corresponding to each side of a regular polygon is an isosceles $\triangle$ with its vertex at the centre, O, of the circumscribing circle as shown in Fig. 103. In the case of the hexagon all these triangles are equilateral.

107. Inscribed Circles.

Circles which are drawn **within** polygons so as to **touch** all the sides are called inscribed circles. They clearly touch each side at its mid-point.

To draw the inscribed circle of a given polygon, such as the hexagon in Fig. 104, find the centre, which is the same as for the circumscribing circle, by drawing the perpendicular bisectors of the sides.

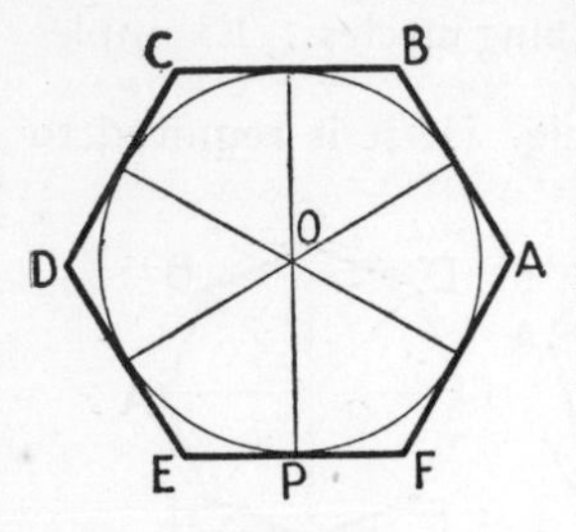

FIG. 104.

The length of any one of these perpendiculars from O to one of these mid-points is the radius of the inscribed circle. For example, in Fig. 104, OP is the radius of the inscribed circle.

108. Angles of Regular Polygons.

As stated in § 105 these are all equal in any regular polygon. They can be found in various ways.

First method. Consider the pentagon in Fig. 103 (*b*). The five isosceles Δs with sides of the pentagon as bases and vertices at O are clearly congruent, and the angles at centre are equal. Their sum is four right angles. This is evidently true for all regular polygons.

For the pentagon each angle at the centre $= \frac{1}{5}$ of 4 right $\angle$s.

$$\therefore \quad \angle AOB = \frac{360^\circ}{5} = 72^\circ.$$

$$\therefore \text{ each of the angles } OAB, OBA = \tfrac{1}{2}(180^\circ - 72^\circ) = 54^\circ.$$

$$\therefore \quad \angle ABC = 2 \times \angle OBA = 108^\circ.$$

Second method. For our example in this case we will consider the hexagon (Fig. 103 (*a*)).

In this polygon six equilateral Δs are formed with sides of the polygon as bases.

$\therefore$ sum of all the angles of these triangles, $6 \times 2 = 12$ right $\angle$s.

These include the angles at O, *i.e.*, 4 right $\angle$s.

$\therefore$ sum of the six angles of the hexagon $= (12 - 4)$ right $\angle$s.

$= 8$ right $\angle$s.

$\therefore$ each angle $= \frac{8}{6}$ right $\angle$s $= 8 \times \frac{90°}{6}$

$= 120°$.

The angle of a regular polygon of n sides.

In general, if a polygon has n sides there are n **triangles.**

$\therefore$ sum of all the angles of the Δs $=$ **2n right angles.**

This includes the 4 right $\angle$s at the centre.

$\therefore$ **sum of angles of the polygon $= (2n - 4)$ right $\angle$s.**

$$\therefore \text{ each angle} = \frac{2n - 4}{n} \text{ right } \angle\text{s.}$$

In the octagon, for example, $n = 8$.

$\therefore$ each angle $= \frac{(2 \times 8) - 4}{8}$ right $\angle$s

$= 135°$.

Third method. Exterior angles of a regular polygon.

Let the sides of a regular polygon (Fig. 105) be produced in the same order.

Exterior angles are thus formed, and **these are as many as there are sides to the polygon.**

Let $n =$ the number of sides.

Then there are n exterior angles. At each angular point the sum of the interior and exterior angles is 2 right $\angle$s.

$\therefore$ for n exterior angles:

sum of **all** the angles $=$ **2n right $\angle$s.**

but sum of **interior** angles $= (2n - 4)$ right $\angle$s (*see above*).

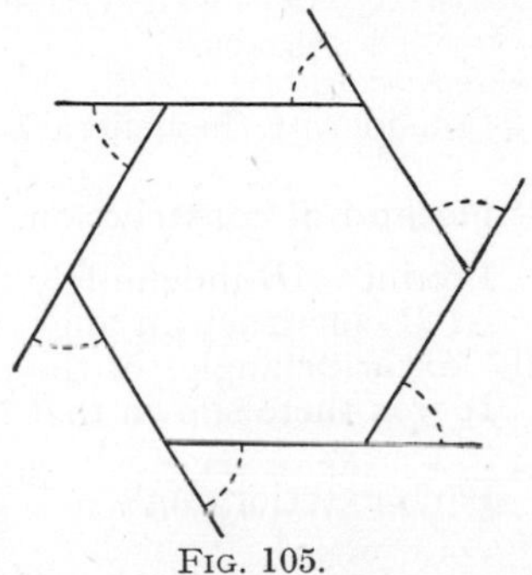

FIG. 105.

$\therefore$ **sum of exterior angles $=$ 4 right $\angle$s.**

When there are n sides and thus n exterior angles:

$$\text{each } \textbf{exterior} \text{ angle} = \frac{4 \text{ rt. } \angle\text{s}}{n} = \frac{360°}{n}.$$

$\therefore$ each **interior** angle $= 180° - \frac{360°}{n}$.

Thus, for a hexagon each interior angle

$$= 180° - \frac{360°}{6} = 120°.$$

109. Construction No. 10.

To construct a regular polygon on a given straight line.

This can readily be effected by making use of the angle properties of polygons which have been stated above. Several methods can be employed—occasionally special methods as in the case of a hexagon—but the following general method can always be used.

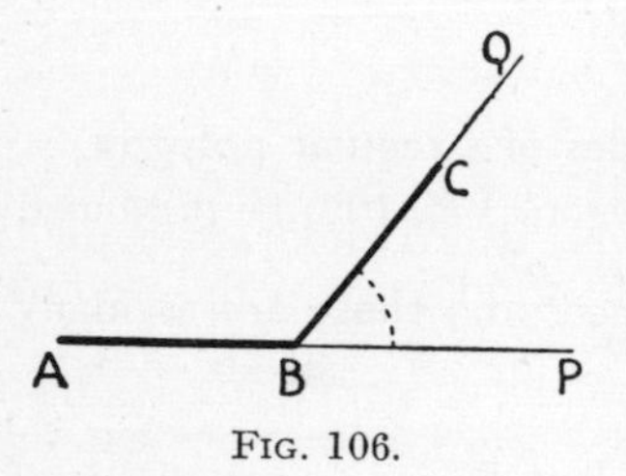

FIG. 106.

Let AB (Fig. 106) be the given side of the required polygon.

Considering the general case let the polygon have n sides.

Method of construction.

Produce AB indefinitely to P.

At B construct an angle which, by calculation, is one of the **exterior** angles of the polygon (§ 108, Third Method). It was there shown that :

Each exterior angle $= \frac{360°}{n}$.

$\therefore$ **at B construct an angle PBQ equal to** $\frac{360°}{n}$.

Along PQ mark off $BC = AB$.

Then **BC is a side of the polygon** and $\angle ABC$ is one of the angles of the polygon.

Similarly, another angle equal to PBC can be constructed

at C and a third side obtained. Thus the whole polygon can be constructed step by step.

Note.—It is frequently helpful, having obtained BC, to find the centre of the circumscribing circle as stated above.

Exercise 11

1. Find the number of degrees in each of the equal angles of the following regular polygons : (*a*) heptagon, (*b*) octagon, (*c*) decagon.

2. If each of the interior angles of a regular polygon is 160°, how many sides are there?

3. Construct a hexagon with a side of 1 in.

   (*a*) What is the radius of the circumscribing circle?
   (*b*) What is the distance between two opposite sides?
   (*c*) Find the area of the hexagon.

4. Construct a regular octagon of side 1·5 in.

5. Each of the exterior angles of a regular polygon is 40°. How many sides has it? Find each of the interior angles.

6. The sum of the exterior angles of a polygon is equal to the sum of the interior angles. How many sides has the polygon?

CHAPTER 15

LOCI

110. Meaning of a Locus.

If a number of points be marked, without any plan, on a sheet of paper, they will not, in general, lie in any regular formation, nor will they form any regular geometrical pattern. If, however, they are placed so as to satisfy a geometrical condition, they will be seen to lie in a recognisable geometric figure.

A simple example is that of drawing a straight line from a point A to another point B, by means of a straight edge or ruler. As the point of the pencil moves along the edge of the ruler it can be regarded as forming a continuous succession of points, the whole constituting a straight line. All such points satisfy the condition of lying on the straight line joining A to B (*see also* § 6).

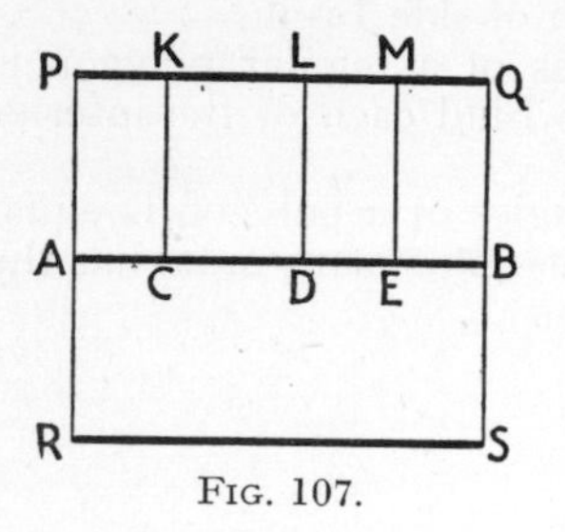

FIG. 107.

A second example is as follows. Let AB, Fig. 107, be a fixed straight line and suppose it is desired to find **all the points in the plane of the paper which are half an inch from it.**

We know that the distance of a point from a straight line is the length of the perpendicular drawn from the point to the line. Take any point C on AB and draw CK, perpendicular to AB and half an inch in length. Then the point K at the end of the line satisfies the condition of being half an inch from AB. Similarly, taking other points on AB, such as D and E and drawing perpendiculars DL, EM, each half an inch in length, we obtain the points L and M, which also satisfy the condition.

It is clear, from previous work, that **all such points must**

lie on the straight line PQ which is parallel to AB and equal in length to it. The points P and Q will lie at the ends of the perpendiculars from A and B respectively.

It is also evident that a straight line RS, drawn parallel to AB on the other side of it, such that the perpendicular from any point on AB to it is half an inch in length, also contains points which satisfy the condition of being half an inch from AB.

Consequently, we conclude that *all the points in the plane which satisfy the condition of being half an inch from* AB *lie on two parallel straight lines, half an inch from* AB, *equal in length to* AB, *and lying on opposite sides of it.*

Further, it will be clear that **there are no other points** in the plane of the paper which are distant half an inch from AB.

An assemblage of all the points which satisfy a given condition is called a locus. (Latin, *locus* = a place, position; plural—loci.)

We have also seen that the straight line PQ may be regarded as the path of a point moving so that it satisfies the condition of being half an inch from AB. Hence the definition:

The path traced out by a point moving so as to satisfy a fixed condition or law is called the locus of the point.

111. Let us next consider this problem:

What is the locus of points in a plane which are half an inch distant from a fixed point?

The answer to this is at once suggested by the definition of a **circle** (§ 21). In that definition we read "*all straight lines drawn from points on the circumference to a fixed point within the curve called the centre are equal*".

Consequently, the answer to the question is that **the locus is the circumference of a circle, whose centre is at the fixed point and whose radius is half an inch.**

112. Locus of Points Not in One Plane.

In the above definitions points and lines were specified as being in one plane. But it will be readily understood that there may be **points not in the plane,** *i.e.*, in space, which also satisfy the given condition.

In the example of the previous section if the lines are no longer restricted to one plane, then it will be seen that *the locus in space of points which are at a given distance from a fixed point* is the *surface of a ball or sphere.* The radius of the sphere is the specified distance from the fixed point which is the centre of the sphere.

Similarly, **the locus in space of points which are a specified distance from a fixed straight line** (as in § 110) is the **interior surface of a cylinder,** such as a jam-jar or tin.

The sphere and cylinder will be more fully dealt with in Chapter 24.

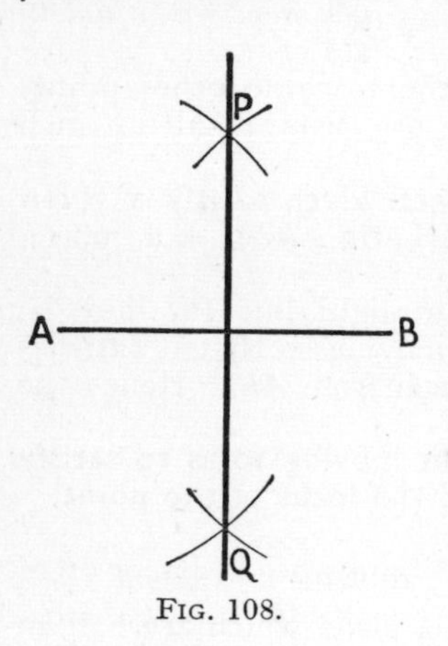

FIG. 108.

113. Locus of Points Equidistant from Two Fixed Points.

Let A and B be the two fixed points (Fig. 108).

Construction No. 5 provides the solution of the problem. There it was proved that any point on the perpendicular bisector of a straight line is equidistant from the ends of the line.

Therefore, join AB and draw as in Construction No. 5 the perpendicular bisector of AB, viz., PQ.

Then, as in the theorem of § 68, any point on PQ is equidistant from A and B.

$\therefore$ **PQ is the required locus.**

114. Locus of Points Equidistant from Two Intersecting Straight Lines.

The answer to this problem is supplied by the Theorem of § 68, following on Construction No. 4.

It is clear that the locus is **the bisector of the angle** formed by the intersection of the two straight lines. Thus, in Fig. 72 *OP* is the locus of points which are equidistant from the two straight lines *OA* and *OB*.

115. The geometrical construction of a locus is seldom as easy as those stated above. The following example is somewhat more difficult.

Find the locus of a point which moves so that the sum of its distances from two fixed points is constant.

The locus may be drawn as follows:

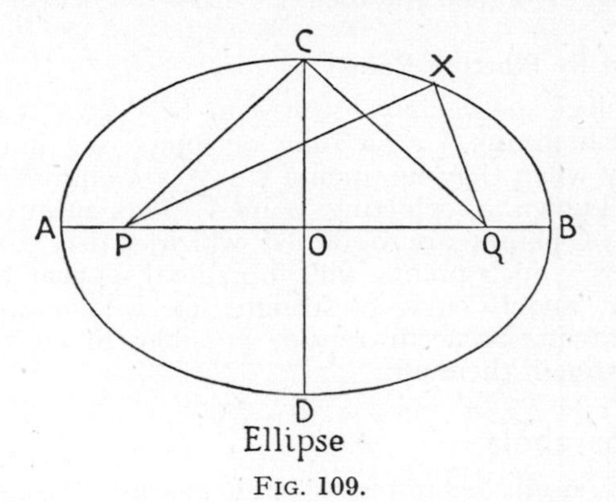

Ellipse

FIG. 109.

Let *P* and *Q* be two fixed points on a piece of paper on a drawing board.

Fasten two pins firmly at *P* and *Q*.

Take a closed loop of fine string or thread and place it round the pins at *P* and *Q*.

With the point of a pencil stretch the string taut, so that it takes up a position such as *X*.

If the pencil now be moved, keeping the string taut, the point at *X* will travel along a curve. Since the length of the string is constant, and the distance between *P* and *Q* is constant, the sum of *PX* and *QX* must be constant.

Thus, the point moves round a curve so that **the sum of its distances from P and Q is constant.**

The resulting curve is an **ellipse,** which may therefore be defined as:

The locus of the point moving so that the sum of its distances from two fixed points is constant is an ellipse.

Two of the points A and B will lie on the straight line PQ produced each way.

Bisect AB at O.

Through O draw COD perpendicular to AB.

Then AB and CD are called the **major and minor axes** of the ellipse.

P and Q are called the **foci.**

116. Loci by Plotting Points.

Many loci, as we have seen, can be drawn readily by mechanical means, *e.g.*, a ruler or compasses, but others, especially when they lie upon a curve, are obtained by the method known as plotting points. This means that a number of points are obtained which satisfy the given conditions. Such points will, in general, appear to lie on a regular, smooth curve or straight line. They are joined up by drawing as accurately as possible the curve which passes through them all.

The parabola.

A very useful example of this method is one which produces a **parabola.** A mechanical method of drawing this curve is possible, but is seldom used. The following two methods are commonly employed.

(1) **Geometrical.** In this method we employ a fundamental property of a parabola as a locus. It is as follows:

The locus of points whose distances from a fixed point are equal to the corresponding distances from a fixed straight line is called a parabola.

This can be drawn most easily by using squared paper as in Fig. 110.

Let P be the fixed point and XOX' the fixed straight line, O being the point where the perpendicular from P to XOX' meets that line.

If OP be bisected at A, then A is clearly a point on the curve, its distance from the fixed straight line XOX', viz., AO, being equal to AP, its distance from P.

Selecting one of the ruled lines perpendicular to OP, such as BC, with BO as radius and P as centre, draw an arc of a circle cutting BC in C.

Then $$PC = OB = CD.$$

I.e., the distance of C from the fixed point P is equal to its distance from the fixed line OY.

$\therefore$ **P is a point on the locus.**

A similar point can be found on the other side of

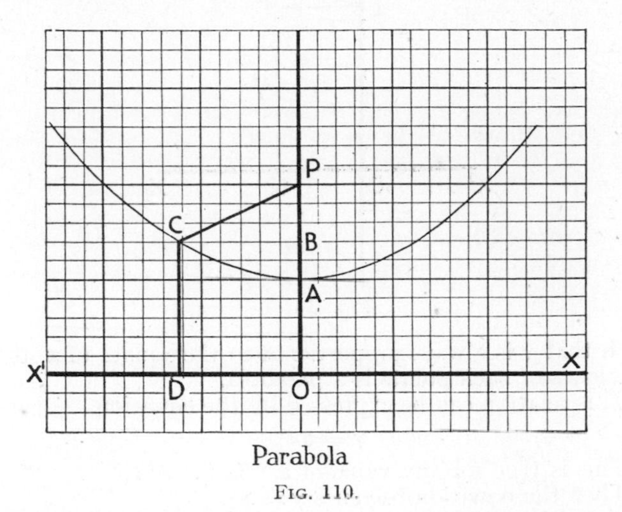

Parabola

FIG. 110.

OP. Thus a number of other points may be found on both sides of OP as an axis and a curve drawn to contain them. The more points that are plotted the more accurately the curve can be drawn.

(2) **Algebraical.**

Students who have studied algebra will be acquainted with the following, which is a very brief summary of the treatment of the matter in an algebra text-book (*Teach Yourself Algebra*, § 108).

Let OX, OY be two straight lines at right angles to each other.

Let P be a point in the plane of these lines.

Let y, *i.e.*, PK, be its distance from OX.

Let x, *i.e.*, PL, be its distance from OY.

If the relation between y and x for a series of points be

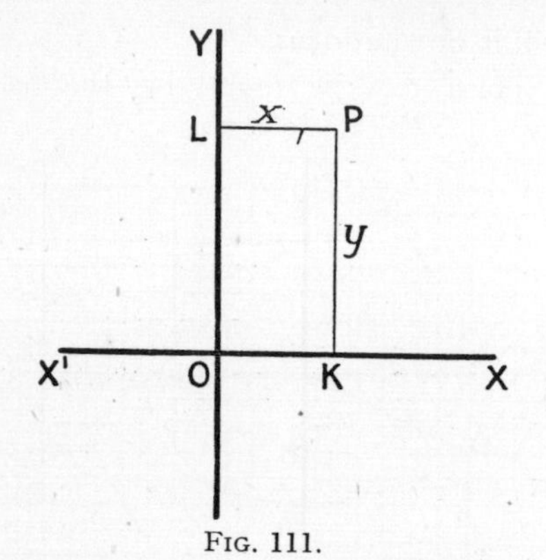

FIG. 111.

such that y is always proportional to the square of x, then the locus of such points is a **parabola**.

This relation can be expressed by the equation

$$y = ax^2.$$

This is true for any value of **a**. Let $a = 1$.

Then the equation becomes $y = x^2$.

Using this simpler form of the equation, we may proceed to find the locus of all points which satisfy the condition.

To do this we assign suitable values to x and then calculate the corresponding values of y. For convenience some of these are tabulated as follows (other values can be added by the student):

x	0	0·5	1	1·5	2	2·5	3
y	0	0·25	1	2·25	4	6·25	9

Using scales as indicated on OX and OY (Fig. 112), we proceed to find the points for which the corresponding values of x and y are those in the table, thus at P

$$x = 2, y = 4.$$

It will be seen that these points apparently lie on a smooth regular curve. This must be drawn by the student. It is a reasonable inference from the form of the curve

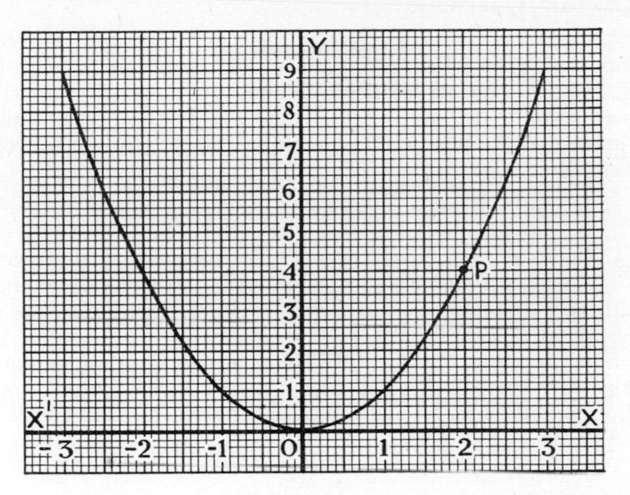

Parabola

FIG. 112.

that all points on it, besides those plotted, will satisfy the condition $y = x^2$. This can be checked by taking points on it, finding the corresponding values of x and y and seeing if they do satisfy the condition. Further, it will be clear that there are no points on the plane, not lying on the curve, which satisfy the condition $y = x^2$.

For convenience, different units are employed for x and y.

The curve is thus the locus of the points, which are such

that the distance from OX is equal to the square of the distance from OY.

The student who has a knowledge of elementary algebra will realise that there is a similar curve on the other side of OY, corresponding to negative values of x. This agrees with the curve as drawn by geometrical methods in Fig. 110. The curve is a **parabola.**

117. The Hyperbola.

Algebraic expressions, involving two quantities denoted by x and y, in which y is expressed in terms of x, can be represented by curves obtained in a similar way to that given above. A noteworthy example is the curve which represents the relation between x and y denoted by the equation

$$y = \frac{1}{x}.$$

Using the method of the previous example, the curve to be obtained may be regarded as the locus of points such that the distance (y) of each of them from OX is the reciprocal of the distance (x) from OY.

This curve presents difficulties when x becomes very large or very small, but they cannot be discussed here. The student is referred to *Teach Yourself Algebra*, § 173.

A table of corresponding values of x and y is as follows:

x	$\frac{1}{4}$	$\frac{1}{2}$	1	2	3	4	
y	4	2	1	$\frac{1}{2}$	$\frac{1}{3}$	$\frac{1}{4}$	

When the curve is drawn through the points obtained from these it is as shown in Fig. 113.

A curve similar to that obtained by using the above values can be drawn for negative values of x.

This curve is known as the **hyperbola.**

118. The Cycloid.

This curve is the locus of a fixed point on the circum-

ference of a circle which rolls along a straight line without slipping.

The curve, which is one of considerable practical value, may be observed by making a visible mark on a bicycle wheel, or garden roller. As the wheel rolls smoothly the

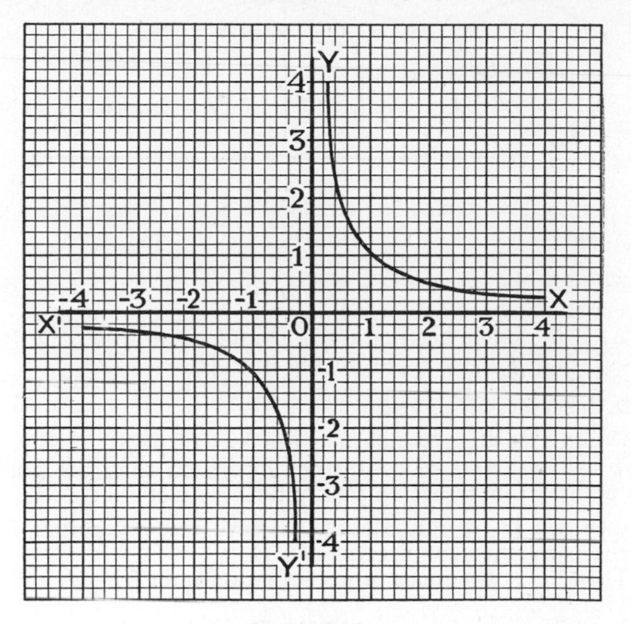

Hyperbola

Fig. 113.

mark will be seen to move along a curve in space. This curve is the **cycloid**.

A cycloid may be plotted as follows. Take a solid circular disc, place it horizontally on a piece of paper and with its edge touching a fixed ruler or a book a marked point, *P*, is made on the paper. Carefully roll the disc along the edge of the ruler for a short distance, taking care it does not slip. Now make a point on the paper corresponding to the new position of P. Repeat this and so obtain a number of similar points; the curve drawn through these is the cycloid.

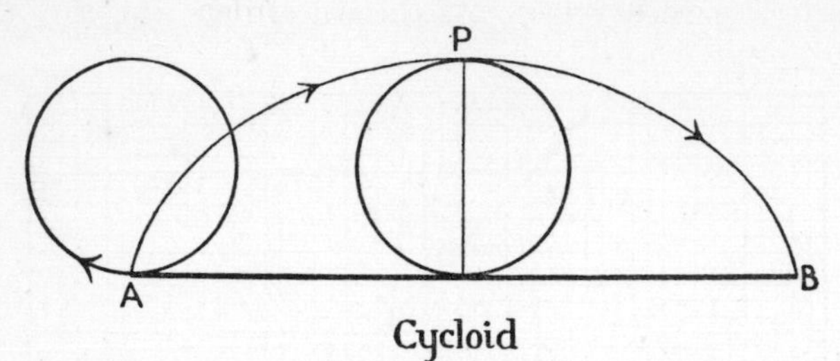

FIG. 114.

Fig. 114 represents the curve, the marked point on the circle starting from A, reaching the highest point at P.

At B the circle has made one complete rotation, and the fixed point is back again on the line AB.

119. Intersection of Loci.

When two lines which are the loci of points satisfying two separate sets of conditions intersect, then the point or points of intersection satisfy both sets of conditions.

Example 1. A and B (Fig. 115) are any two points, PQ is the perpendicular bisector of AB.

Then PQ is the locus of points equidistant from A and B (§ 113).

Let C be a third point, and RS the perpendicular bisector of BC.

Then RS is the locus of points equidistant from B and C.

Let O be the intersection of PQ and RS.

Since O lies on PQ it must be equidistant from A and B.

Also since O lies on RS it must be equidistant from B and C.

∴ O must satisfy both sets of conditions and is equidistant from A, B and C,

i.e., $$OA = OB = OC.$$

∴ if a circle be described with O as centre and OA as radius the circumference will pass through B and C.

The following conclusions may be deduced from the above:

(1) *If* AC *be drawn then* ABC *is a triangle and the circle*

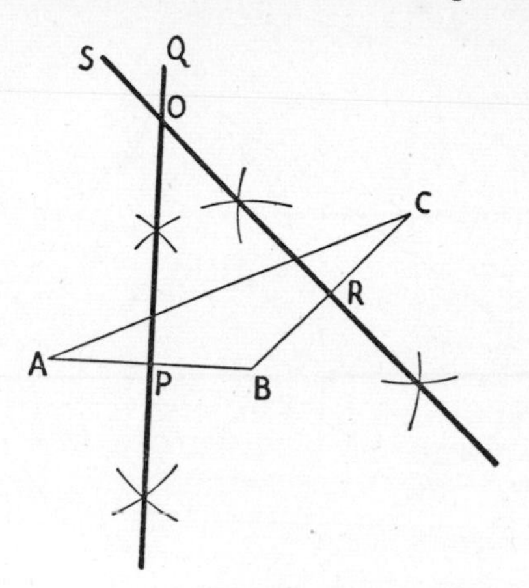

Fig. 115.

drawn as described above is the circumscribing circle of the triangle (see § 106).

(2) *Since PQ and RS can intersect in one point only, one circle only can be described to pass through three points.*

(3) *The perpendicular bisector of AC must pass through the centre of the circumscribing circle O. Consequently the perpendicular bisectors of the sides of a triangle must be concurrent.*

Example 2. The principle of the intersection of loci has been used previously in a number of examples, without reference to a locus, as, for example, in the following problem.

A *and* B *are two points* 2·5 *in. apart on a straight line* AB. *Find a point which is* 2 *in. from* A *and* 1·5 *in. from* B.

With centre *A* and radius 2 in. describe a circle (Fig. 116).

With centre *B* and radius 1·5 in. describe a circle.

The locus of all points in the plane which are 2 in. from A is the circumference of a circle, centre A and radius 2 in.

The locus of all points 1·5 in. from B is the circumference of a circle centre B and radius 1·5 in.

The intersection of these two circles, viz., C and C' are

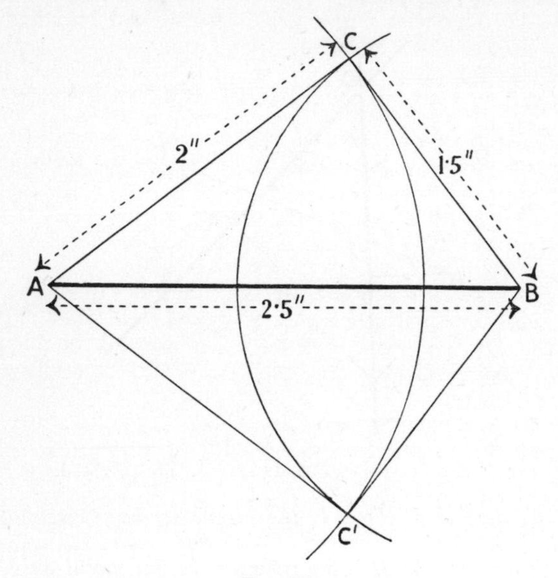

Fig. 116.

points which satisfy both conditions, *i.e.*, each of them is 2 in. from A and 1·5 in. from B. They are thus the vertices of two triangles ABC and ABC′ whose sides are of the given dimensions, 2·5 in., 2 in. and 1·5 in.

These $\triangle$s are clearly congruent.

The student will observe that, in principle, this was the method employed in the construction of a triangle, when three sides are given (*see* § 45, B).

Exercise 12

1. Describe the following loci:

(*a*) The centre of a wheel of radius 1 ft. which rolls in a vertical plane over a smooth horizontal surface.

(*b*) The centre of a wheel of radius 1 ft. which rolls round a wheel of radius 2 ft.

(*c*) A runner who runs round a circular track, always keeping 2 yds. from the inner edge of the track.

2. A number of triangles on the same base and on the same side of it are of equal area. What is the locus of their vertices?

3. On a given straight line as base a number of isosceles triangles are constructed. What is the locus of their vertices?

4. On a given straight line, AB, a number of right-angled triangles are constructed, each with the right angle opposite to AB. Draw a number of such $\triangle$s and sketch the curve which passes through the vertices. What does it appear to be?

5. AB is a fixed straight line and O a point without it. O is joined to a point P on AB and PO is produced to Q so $OQ = OP$. As P moves along AB, what is the locus of Q?

6. On a fixed straight line, AB, a series of isosceles $\triangle$s are constructed on one side of AB. Let C be one of the vertices. Produce CA to D, so that $CD = CB$. What is the locus of D?

7. AB is a straight line and P is a point without it. If P moves so that the perpendicular from it to AB is always one-half of its distance from A, what is the locus of P?

8. XY is a fixed straight line of indeterminate length. A part of it, BC, is the base of an equilateral $\triangle\ ABC$. If the triangle rolls over, without slipping, on XY, until AC lies on XY, what is the locus described by B?

9. Two straight lines PQ and XY of unlimited length intersect at O at an angle of $45°$. Show how to find points which are half an inch from each of the straight lines.

10. POQ is an angle of $60°$. Show how to find a point which is half an inch from OP and 2 in. from OQ.

CHAPTER 16

THE CIRCLE. ARCS, CIRCUMFERENCE, AREA

120. Arcs and the angles they subtend.

The student is reminded of the conclusions reached in §§21 and 22. There it was pointed out that if a straight line rotates in a plane about a fixed point at one end of the line then any point on the rotating line traces the **circumference** of a circle, and that any part of this circumference is called an **arc**.

Thus when the straight line OA (Fig. 117) rotates to OB about O, an arc of the circle, viz., AB is described by A, and $\angle AOB$ is the corresponding angle through which AB turns.

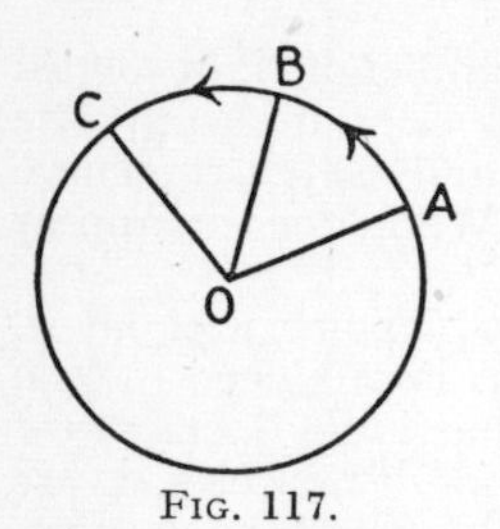

Fig. 117.

The angle AOB is said to stand on the arc AB, while the **arc AB is said to subtend the angle AOB at the centre of the circle.** Both arc and angle are described by the same amount of rotation.

If the rotating line moves through a **further angle BOC equal to AOB**, the arc BC is formed subtending $\angle BOC$ at the centre. Clearly since $\angle AOB = \angle BOC$ the arc BC must equal the arc AB. It is reasonable to conclude that **equal angles correspond to equal arcs** and vice versa. This may be expressed in the theorems.

Theorems.

(1) Equal arcs in a circle subtend equal angles at the centre.

(2) (*The converse of the previous theorem.*) Equal angles at the centre of a circle stand upon equal arcs.

When, as stated above and illustrated in Fig. 117, the

amount of rotation, and therefore the angle described, is doubled, the arc is also doubled. If the angle, in the same way, were to be trebled, the arc would be trebled, and so for other multiples. It may therefore be concluded that:

In a circle arcs are proportional to the angles which they subtend at the centre of the circle.

121. Sector.

That part of a circle which is enclosed by an arc and the two radii drawn to the extremities of the arc is called a sector.

In Fig. 118 the figure *AOB* is a **sector.** In a circle of

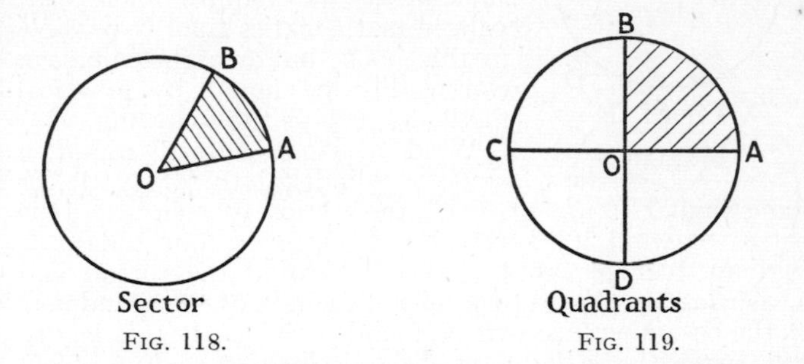

Fig. 118.

Fig. 119.

given radius, the size of the sector is determined by the angle of the sector, *AOB*, or by the length of the arc.

Quadrant. If the angle of the sector is a right angle, the sector becomes a **quadrant.**

In Fig. 119 the shaded sector is a quadrant.

If two diameters be drawn at right angles, such as *AC* and *BD*, the circle is divided into **four quadrants.**

Semicircle. If the angle of the sector be 180°, the sector becomes a semi-circle, *i.e.*, half a circle, as *ABC* in Fig. 119. A semi-circle thus contains two quadrants. An important practical example is the semi-circular protractor (*see* Fig. 21).

Chord. The straight line which joins the ends of an

arc of a circle is called a chord. In Fig. 120, *AC* is a chord of the circle in which *ABC* is an arc. It is also a chord of the arc *ADC*.

A **diameter** is a chord which passes through the centre.

122. Length of the circumference of a circle.

The length of a curve obviously cannot be obtained in the same way as that of a straight line by means of a straight edge or ruler. Hence, other methods must be found.

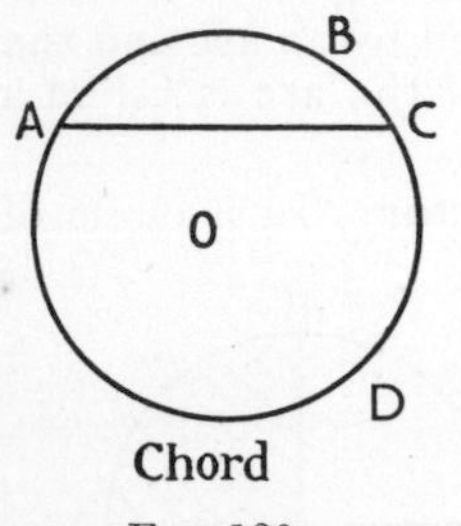

Fig. 120.

The length of the circumference of a circle is a matter of great importance, if only for practical purposes. To obtain an exact formula for its calculation we require more advanced mathematics than is possible in this book, but an approximation can readily be found by practical methods, such as the following.

Wind a stout thread round a smooth cylindrical tin or bottle, or some similar object of which the section is a circle. It is better to wind it exactly round three or four times, so that an average can be taken. Unwind the thread and measuring the thread by a ruler the length of one round, *i.e.*, of the circumference can be found. Evidently this length will vary with the diameter of the circle, so the next problem will be to find the relation between them. The diameter can be measured by means of a ruler, care being taken to see that the line measured passes through the centre of the circle. This can be done more satisfactorily in the case of a jar or tin by the use of callipers, or by placing the circular object on its side on the table between two smooth rectangular blocks, taking care that they are parallel. The distance between the blocks is evidently the diameter of the circle.

From these measurements the **ratio of the length of the circumference to that of the diameter** may be found. It is better to do this with several objects of varying diameters, and then take the average of the results. Two conclusions will be apparent:

(1) The value of the ratio $\frac{\text{circumference}}{\text{diameter}}$, allowing for errors of measurement, is found to be the same in all cases.

If $\quad C$ = circumference
and $\quad d$ = diameter.

Then $\quad \frac{C}{d}$ is a constant number.

(2) This constant number will probably be found by the above experiment to lie between 3·1 and 3·2.

If this constant ratio can be determined accurately we have a rule by which the length of any circumference can be found when the diameter is known.

The problem of finding the ratio exactly has exercised mathematicians for many centuries. The Egyptians arrived at fairly good approximations and the Greeks at more exact ones. Modern mathematics has, however, found a method by which it can be calculated to any required degree of accuracy. Its value to 7 places of decimals is 3·1415927... This is universally denoted by the Greek letter π (pronounced " pie ").

Thus $\quad \pi = 3{\cdot}1415927\ldots$ to 7 places.

For practical purposes it is usually sufficient to take

$$\pi = 3{\cdot}1416.$$

A less accurate value is $3\frac{1}{7}$, *i.e.*, $\frac{22}{7}$.

Since $\frac{22}{7} = 3{\cdot}1428\ldots$ it is clear that if it be so used in calculations the accuracy of the results cannot, in general, be depended upon as accurate for more than **two significant figures**.

Using this symbol, the results reached above can be expressed in a formula:

Let $\quad C$ = length of circumference.
$\quad d$ = length of diameter.
$\quad r$ = length of radius.

Then
$$\frac{C}{d} = \pi$$

or
$$C = \pi d$$

or since $$d = 2r$$
$$C = 2\pi r.$$

123. To Find the Length of an Arc.

The length of an arc of a circle, given the angle subtended by it at the centre, can readily be calculated from that of the whole circumference by making use of the geometric theorem of § 120, viz.:

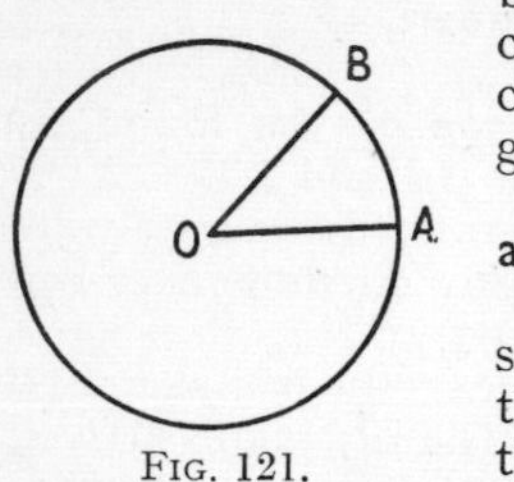

FIG. 121.

Arcs are proportional to the angles they subtend at the centre.

In Fig. 121 the arc AB bears the same ratio to the whole circumference that the angle subtended by AB at the centre, viz., AOB, bears to a complete rotation, 360°.

Let a = length of arc.
r = radius of circle.
n° = angle subtended at the centre by AB in degrees.

Then as stated above:

$$\frac{a}{2\pi r} = \frac{n^\circ}{360^\circ}.$$

$$\therefore \quad a = \frac{n}{360} \times 2\pi r.$$

For example, if an arc subtends 72° at the centre of a circle of radius 2 in.:

$$\text{Then, length of arc} = \frac{72}{360} \times 2\pi \times 2$$
$$= \frac{4\pi}{5} = \frac{4 \times 3{\cdot}1416}{5}$$
$$\doteq 2{\cdot}51 \text{ in. approx.}$$

124. The Area of a Circle.

Areas bounded by regular curves are, in general, not easily calculated, except by methods of advanced mathematics. The problem of finding the area of a circle is no exception to this. It was one of the most famous of geometrical problems for centuries, but was never solved satisfactorily until modern mathematics found means of obtaining the area of any figure bounded by a regular curve. This method is beyond the scope of the present volume.

As was the case with the circumference, however, there are methods by which the area can be determined approximately, and one of these is given below.

In the circle drawn in Fig. 122, AB, BC, CD . . . are

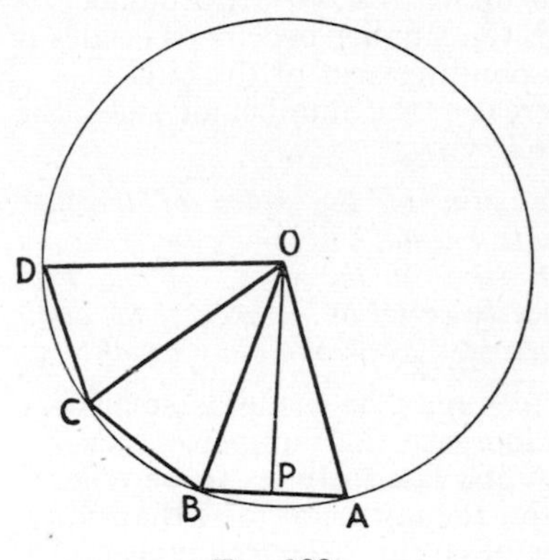

FIG. 122.

the sides of a regular polygon inscribed in the circle; centre O, and radius r. The arcs corresponding to these sides are equal. The angular points A, B, C . . . are joined to the centre, thus forming a series of equal isosceles triangles, OAB, OBC, OCD . . . (§ 106).

Selecting one of these $\triangle$s, OAB, draw OP perpendicular to AB. This is the altitude or height of the triangle, and in each triangle there is a corresponding equal altitude.

Let $\qquad OP = h.$

Then $\qquad$ area of $\triangle OAB = \frac{1}{2}AB \times h$

$\qquad$ area of $\triangle OBC = \frac{1}{2}BC \times h,$

and $\qquad$ area of $\triangle OCD = \frac{1}{2}CD \times h.$

Similarly for all such $\triangle$s corresponding to other sides of the polygon.

Taking the sum of these areas

$$\triangle OAB + \triangle OBC + \triangle OCD + \ldots$$
$$= \tfrac{1}{2}(AB + BC + CD \ldots) \times h$$
$$= \tfrac{1}{2}(\text{perimeter of polygon}) \times h \quad . \quad (A)$$

Suppose the number of sides of the polygon to be greatly increased. Then the number of isosceles triangles is increased correspondingly. The result (A) above remains true, however many may be the number of sides. The more they are, the smaller become the sides of the polygon and the corresponding arcs of the circle.

Thus, the greater the number of sides, the more nearly true are the following:

(1) *The sum of the sides of the polygon becomes approximately equal to the circumference of the circle.*

(2) *The sum of the areas of the triangles becomes approximately equal to the area of the circle.*

(3) h *becomes approximately equal to* r.

The differences may be made as small as we choose, by continuing to increase the number of sides.

But result A above continues to be true.

$\therefore$ ultimately, it may be argued that, approximately,

$$\textbf{Area of circle} = \tfrac{1}{2}(\textbf{circumference}) \times \mathbf{r}.$$

But circumference $= 2\pi r$.

$\therefore$ **area of circle** $= \frac{1}{2} \times 2\pi r \times r$,

or
$$\textbf{Area} = \pi r^2.$$

This formula may also be expressed in terms of the diameter.

If
$$d = \text{diameter}$$
$$d = 2r.$$
$$\therefore \quad r = \frac{d}{2}.$$

Substituting in the formula above:

$$\textbf{Area} = \frac{\pi d^2}{4}.$$

125. Area of a Sector of a Circle.

As in the case of the length of an arc, the area of a sector is proportional to the angle subtended at the centre by the arc.

$$\therefore \quad \text{if } x^\circ = \text{angle of the sector,}$$

$$\text{Area of sector} = \frac{x^\circ}{360^\circ} \times \pi r^2.$$

Exercise 13

1. In a circle of 5-in. radius, find the lengths of:

    (*a*) the circumference;
    (*b*) the arc of a quadrant;
    (*c*) an arc subtending an angle of 60°;
    (*d*) an arc subtending an angle of 45° ($\pi = 3{\cdot}1416$).

2. A circle of radius 4 in. passes through the vertices of an equilateral triangle. Find the lengths of each of the arcs opposite to the sides.

3. Find the circumference of a garden roller, diameter 2 ft. 3 in. ($\pi = \frac{22}{7}$).

4. The diameter of a halfpenny is exactly one inch. Find (1) the length of its circumference, (2) its area ($\pi = 3{\cdot}1416$).

5. In a circle of radius 5 in. find the lengths of the arcs which subtend at the centre the following angles: 30°, 110°, 125°.

6. Through what distance does a point at the end of the minute hand of a clock, 3·7 in. long, move between five minutes past three and a quarter to four.

7. A pendulum, consisting of a small leaden bob, at the end of a piece of cotton $4\frac{1}{2}$ ft. long, swings 25° on each side of the vertical. What is the length of the path traced out by the bob on a single swing ($\pi = \frac{22}{7}$)?

8. Find the length of the circumference of a circle the area of which is the same as that of a square of 3 in. side.

9. It was required that the area of the ground covered by the circular base of a tent should be 100 sq. ft. What must be the diameter of the base?

10. A wire 15 in. long is bent round to form a circle. What is the area of the circle?

11. Find the areas of the following sectors:

(*a*) Radius 3 in., angle of sector 60°;
(*b*) Radius 2·8 in., angle of sector 25°;
(*c*) Radius 9 cm., angle of sector 140°;
(*d*) Radius 2·2 in., angle of sector 240°.

12. A searchlight a little above the level of the water of a harbour can turn its rays through an angle of 210°. If the greatest distance at which objects can be clearly seen by the help of it is 1000 yds., what is the area of the surface of the water within which objects can be seen?

CHAPTER 17

CHORDS AND SEGMENTS

126. Chord and Segment.

A chord of a circle has been defined in § 121. It may also be described thus:

If a straight line cuts a circle, that part of it which lies within the circle is called a chord of the circle.

A chord divides a circle into two parts, which are called **segments.**

In Fig. 123 the chord *AB* divides the circle into the two segments *APB*, *AQB*. Unless the chord is a diameter one of the segments is greater than a semi-circle, and is called a **major segment** as *AQB* in Fig. 123.

The other is less than a semicircle and is called a minor segment, as *APB*.

The arcs corresponding may be described as **major and minor arcs.**

The following theorems concerning chords are of considerable importance.

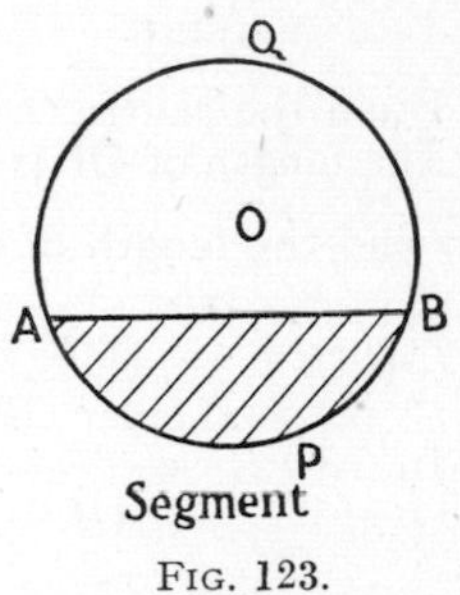

FIG. 123.

127. Theorem. The perpendicular bisector of a chord of a circle passes through the centre.

In Fig. 124 *AB* is a chord of the circle *APB*, *D* is the centre of the chord and *PQ* is perpendicular to *AB*.

Then it is required to prove that *PQ* must pass through the centre *O*.

Proof. *PQ* being the perpendicular bisector of *AB*. It must be the locus of all points equidistant from *A* and *B* (§ 113).

But $OA = OB$ since O is the centre.
$\therefore$ O must lie on PQ.

128. To find the distance of a chord from the centre of a circle.

In the circle ABC (Fig. 125) AB is a chord of known length.

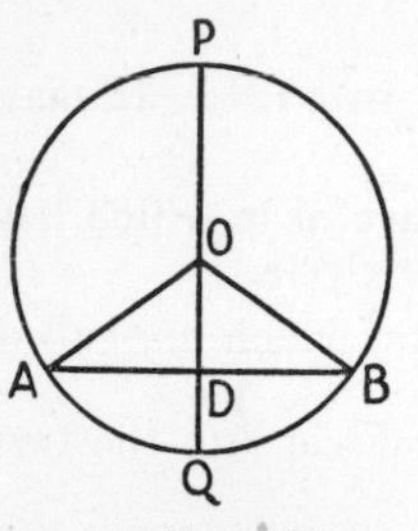

FIG. 124.

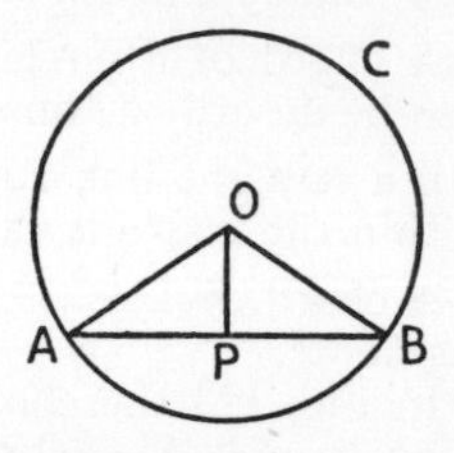

FIG. 125.

From the centre O draw OP perpendicular to AB.
The length of OP is the distance of the chord AB from O.

To find the length of OP.

Join OA, OB.
$\triangle OAB$ is isosceles.
$\therefore$ perpendicular OP bisects the base AB (§ 62 *cor.* 3).
In $\triangle OPB$,

$$OP^2 + PB^2 = OB^2 \text{ (Pythagoras).}$$

$$\therefore \quad OP^2 = OB^2 - PB^2 \quad . \quad . \quad . \quad (A)$$

Whence OP can be found.

Employing algebraic symbols.

Let $2l =$ length of chord.
Then $l =$ length of PB.
Let $h = OP$.
$r =$ radius of circle.

Substituting in (A)

$$h^2 + l^2 = r^2 \quad . \quad . \quad . \quad . \quad . \quad . \quad (B)$$

$$h^2 = r^2 - l^2$$

$$\therefore \quad h = \sqrt{r^2 - l^2}.$$

129. Results (A) and (B) above lead directly to the following theorems.

Theorem. Equal chords in a circle are equidistant from the centre.

In the circle ABC (Fig. 126), AB and DE are equal chords.

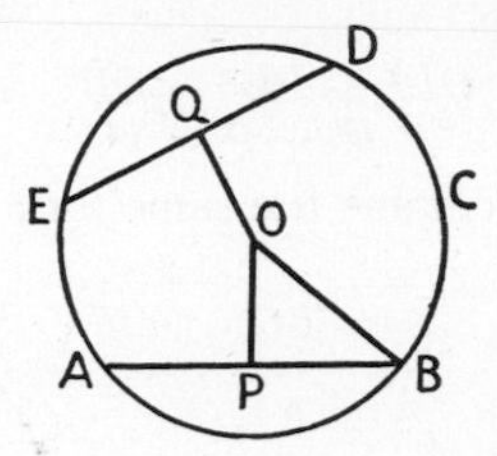

Fig. 126.

Required to prove: The distances from the centre, OP and OQ, are equal.

From (A) in § 128

$$OP^2 + PB^2 = OB^2$$

or from (B) in § 128

$$h^2 + l^2 = r^2$$

(using the same letters as in § 128).

Now PB or l is fixed, and OB or r is fixed.

$\therefore$ OP or h must be fixed, wherever the equal chords are drawn, *i.e.*, $OP = OQ$ and the two chords are equidistant from O.

The converse theorem is obviously true, viz.:

Theorem. Chords of a circle which are equidistant from the centre are equal.

130. From Results (A) and (B) of § 128 the following theorem is readily deduced.

Theorem. The greater of two chords in a circle is nearer to the centre than the lesser.

In the circle ABC (Fig. 127) the chord DE is greater than AB. OQ and OP are the corresponding distances from the centre.

Required to prove:

$$OP > OQ.$$

Proof. As before,

$$OP^2 + PB^2 = OB^2 \quad . \quad . \quad . \quad . \quad . \quad \text{(A)}$$

or

$$h^2 + l^2 = r^2 . \quad . \quad . \quad . \quad . \quad . \quad \text{(B)}$$

In this result r remains the same for all chords.

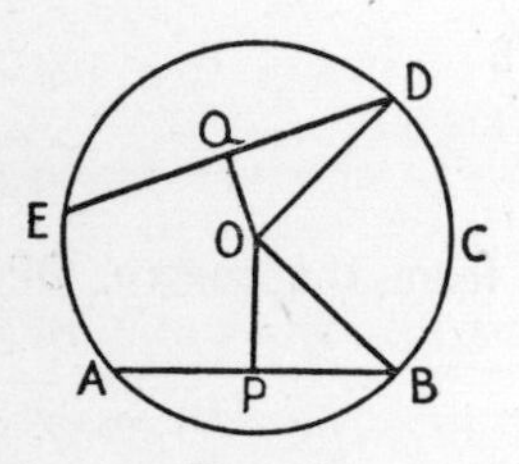

Fig. 127.

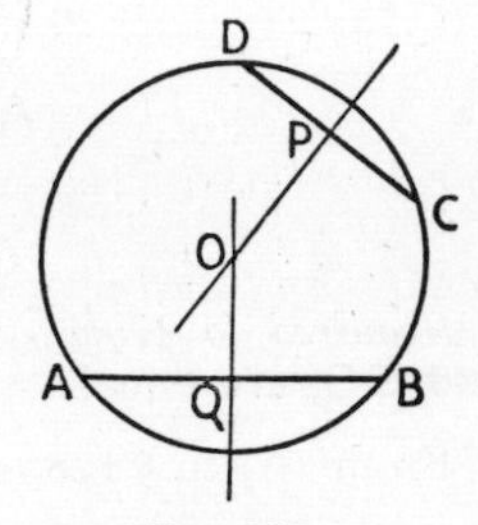

Fig. 128.

$\therefore$ in equation (B), if h be increased, l will diminish, and vice versa.

$\therefore$ for the chords AB and DE,

since

$$QD > PB$$

$$\therefore \quad OQ < OP,$$

i.e., DE is nearer to the centre than AB.

The converse of this also follows from similar reasoning to the above.

131. Construction No. 11.

To find the centre of a given circle.

$ABCD$ (Fig. 128) is a circle of which it is required to find the centre.

Construction. Draw any two chords, AB and CD.

Draw the perpendicular bisectors of each chord, viz., OP and OQ, intersecting at O.

O is the required centre of the circle.

Proof. The perpendicular bisector of AB passes through the centre (§ 127).

Also the perpendicular bisector of CD passes through the centre.

∴ the point which is common to both, viz., O must be the centre.

Exercise 14

1. The diameter of a circle is 5 in. long. How far from the centre is a chord which is 4 in. long?

2. A chord of a circle is 8 in. long and it is 3 in. from the centre. What is the length of the diameter?

3. In a circle whose radius is 13 cm., a chord is drawn 5 cm. from the centre. Find the length of the chord.

4. Find the distance between two parallel chords of a circle which are 24 in. and 10 in. in length. The radius of the circle is 13 in.

5. A is a point on the circumference of a circle centre O. Two equal chords AB and AC are drawn. OA is joined. Prove that OA bisects the angle BAC.

6. A straight line cuts across the circumferences of two concentric circles (§ 21). XY is the chord so formed of the larger circle, and AB is the chord of the smaller circle. Prove that $XA = BY$.

7. In a circle of radius 5 in. a number of chords of length 6 in. are drawn. Find the locus of their mid-points.

8. AB and XY, are parallel chords in a circle. Show that the arc AX equals the arc BY.

9. Draw a circle round a penny and find its centre. Measure its diameter.

10. Draw an equilateral triangle of 2 in. side. Draw the circumscribing circle and measure its radius.

CHAPTER 18

ANGLES IN SEGMENTS

132. Angle in a Segment.

On the arc of the major segment of the circle in Fig. 129, any point C is taken and joined to A and B, the points in which the chord of the segment meets the circumference.

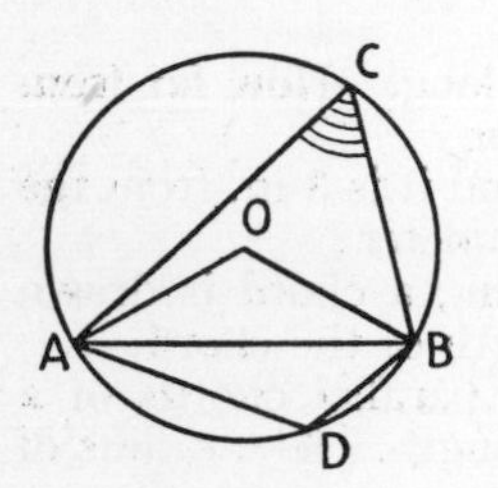

FIG. 129.

Then **the angle ACB is called the angle in the segment.** It is said to be subtended by the chord AB.

Similarly, if a point D be taken in the minor segment the angle ADB is the angle in that segment.

It will be observed that the angle in a **major** segment is an **acute** angle, while the angle in the **minor** segment is **obtuse.**

The angle ACB is also called **the angle subtended at the circumference** by the arc ADB, while the angle AOB is called the **angle subtended at the centre** by the arc ADB.

There is a very important relation between these angles which is expressed in the following theorem.

133. Theorem. The angle which an arc of a circle subtends at the centre is twice that subtended at any point on the remaining part of the circumference.

There are two cases: (1) If the centre O lies within the angle APB (as Fig. 130) and (2) if O lies without the angle as Fig. 131.

Construction. In each case join PO and produce it to meet the circumference again in Q.

Proof. **1st case.**

In $\triangle OAP$, $\quad OA = OP$.

$\therefore \quad \angle OAP = \angle OPA$.

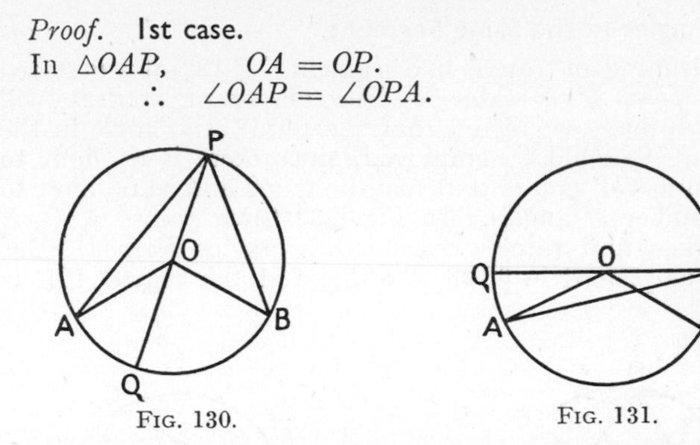

FIG. 130. FIG. 131.

But exterior $\angle AOQ$ = sum of interior $\angle$s OAP, OPA (§ 60).

$\therefore \quad \angle AOQ$ = twice $\angle OPA$.

Similarly from the $\triangle OBP$.

$\angle BOQ$ = twice $\angle OPB$.

$\therefore \quad \angle AOQ + \angle BOQ$ = twice $(\angle OPA + \angle OPB)$,

i.e., $\quad$ **$\angle$AOB = twice $\angle$APB.**

2nd case (Fig. 131).

With the same reasoning as above:

$\angle QOB$ = twice $\angle QPB$,

and $\quad \angle QOA$ = twice $\angle QPA$.

Subtracting **$\angle$AOB = twice $\angle$APB.**

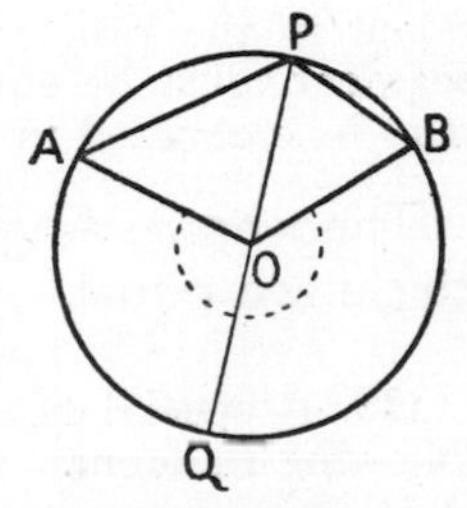

FIG. 132.

A **3rd case** arises when the angle APB is an **obtuse** angle, *i.e.*, it is an angle in a minor segment and stands on a major arc. The angle at the centre, AOB, is now a **reflex** angle.

The proof is similar to the foregoing.

Joining P to O and producing to Q, it is proved, as before, that:

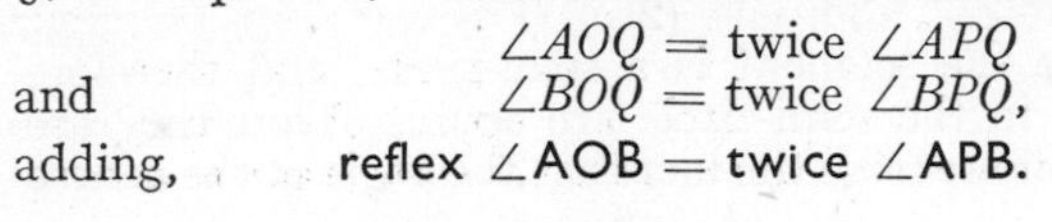

$\angle AOQ$ = twice $\angle APQ$

and $\quad \angle BOQ$ = twice $\angle BPQ$,

adding, $\quad$ **reflex $\angle$AOB = twice $\angle$APB.**

134. Angles in the Same Segment.

In defining an "angle in a segment" (§ 132) it was stated "*any point* C *is taken*". The observant student will probably have wondered that the term "the angle in the segment" should be employed, since there is no limit to the number of points that may be taken and so no limit to the number of angles. In Fig. 133 three points *C, D, E* are taken and three corresponding angles formed. But since, as proved in § 133, the angle at the centre *AOB* is

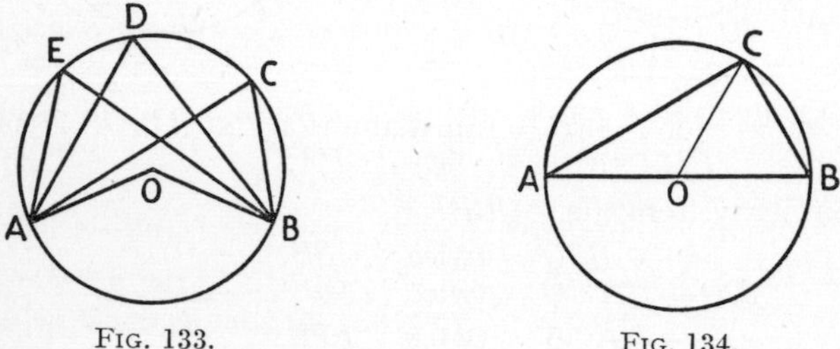

Fig. 133.

Fig. 134.

double **any** angle in the segment, because any angle was taken in the proof, it follows that **all the angles in the segment must be equal.** This striking and important fact may be embodied in a theorem, as follows:

Theorem. Angles in the same segment of a circle are equal.

135. A special case of this theorem is contained in the following theorem.

Theorem. The angle in a semi-circle is a right angle.

In this case the segment is a semi-circle, and therefore all the angles in the semi-circle are equal. That they are right angles follows from the fact that the **angle at the centre** in this case, AOB, Fig. 134, is a "straight" angle, *i.e.*, **equal to two right angles.**

Hence, the angle ACB, being a half of this, is a right angle.

This theorem can also be proved very simply as follows:

Join OC.

Then $\quad OA = OC \quad \therefore \quad \angle OAC = \angle OCA$

also $\quad OB = OC. \quad \therefore \quad \angle OBC = \angle OCB.$

$$\therefore \quad \angle OAC + \angle OBC = \angle ACB.$$

But since the sum of the angles of a triangle is two right angles.

$\therefore$ **$\angle$ACB must be a right angle.**

Note.—This theorem should be compared with the explanation of "an angle in a segment" (§ 132).

136. Quadrilateral inscribed in a circle. The following very important theorem also is easily proved by the Theorem of § 133.

Theorem. The sum of the opposite angles of a quadrilateral inscribed in a circle is equal to two right angles, *i.e.*, the opposite angles are supplementary.

In Fig. 135, $ABCD$ is a quadrilateral inscribed in a circle.

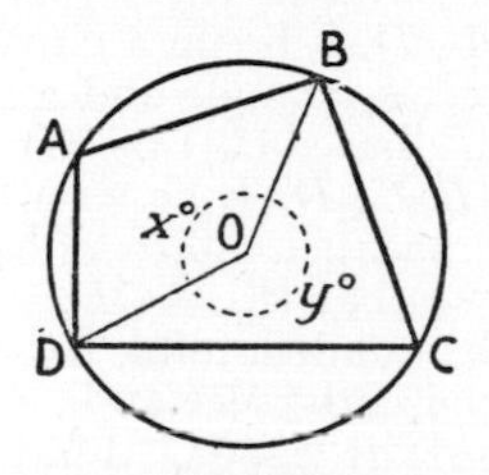

Fig. 135.

Then $\quad \angle A + \angle C = 2$ right $\angle$s

and $\quad \angle B + \angle D = 2$ right $\angle$s.

Proof. Join O to B and D.

$$\angle BOD(x^\circ) = 2\angle BCD, \quad (\S\ 133)$$

and $\quad$ reflex $\angle BOD(y^\circ) = 2\angle BAD. \quad (\S\ 133)$

$\therefore \quad \angle BOD + \text{reflex } \angle BOD = \text{twice } (\angle BCD + \angle BAD),$
but $\angle BOD(x°) + \text{reflex } \angle BOD(y°) = 4$ right angles.

$$\therefore \quad \angle BCD + \angle BAD = \text{two right angles.}$$

Similarly by joining O to A and C it may be shown

$$\angle B + \angle D = 2 \text{ right angles.}$$

Exercise 15

1. In a circle of 2 in., radius cut off a segment which contains an angle of 40°.

2. On a straight line 2 in. long describe a segment of a circle which shall contain an angle of 60°. What is the length of the radius of the circle?

3. A triangle ABC is inscribed in a circle, centre O. The $\angle AOB = 90°$, $\angle AOC = 120°$. Find the angles of the triangle.

4. On a straight line AB, 2 in. long, construct a right-angled triangle of which AB is the hypotenuse and one of the other sides is 0·8 in.

5. In a circle of 1·5 in. radius inscribe a triangle the angles of which are 60°, 40°, 80°.

6. Two triangles ABC, ABD are on the same base and on the same side of it. In the first triangle the angles at the base are 64° and 58°, and in the second triangle 50° and 72°. Show that A, B, D, C lie on a circle.

7. ABC is an isosceles triangle and a straight line DE is drawn parallel to the base, cutting the equal sides in D and E. Prove that B, C, D, E lie on a circle.

8. AB and CD are parallel chords of a circle AD and BC intersect at O. Prove that $OC = OD$.

9. $ABCD$ is a cyclic quadrilateral, *i.e.*, it is inscribed in a circle, and its diagonals intersect at P. If $\angle BPC = 105°$, $\angle BAC = 40°$ and $\angle ADB = 30°$. Find $\angle BCD$.

CHAPTER 19

TANGENTS TO A CIRCLE

137. Meaning of a Tangent.

In Fig. 136, the circle ABC, centre O, is cut by the chord AB. OP is the perpendicular bisector of the chord. Sup-

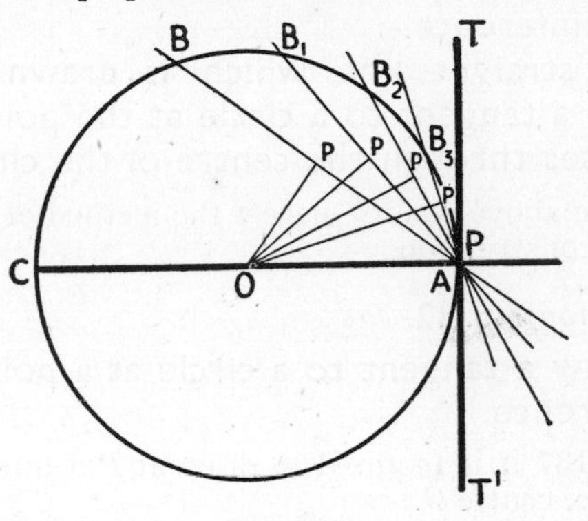

FIG. 136.

pose the chord AB to rotate in a clockwise direction about A as a centre of rotation.

As it rotates the point of intersection B moves round the circumference to a second position B_1 and thus is a shorter distance from A.

At the same time the perpendicular bisector OP rotates, the point P also approaching A.

As the rotation continues the points P and B approach closer and closer to A as shown in the positions B_2, B_3.

Ultimately the point B will move to coincidence with A. Then the straight line AB no longer cuts the circumference in two points. These points coincide and **AB now touches the circle without cutting it,** taking up the position TAT^1.

A straight line which thus meets the circle at one point but, being produced in either direction, does not meet it again, is called a tangent to the circle.

In this final position P also coincides with A, and OP which has throughout been perpendicular to AB, is now perpendicular to AT, the tangent.

These conclusions can be embodied in the following:

(1) A tangent is perpendicular to the radius drawn to the point of contact of the tangent with the circumference.

(2) A straight line which is drawn at right angles to a tangent to a circle at the point of contact, passes through the centre of the circle.

138. The above results suggest the method of solving the following construction.

Construction No. 12.

To draw a tangent to a circle at a point on the circumference.

In Fig. 137 it is required to draw at P a tangent to the given circle, centre O.

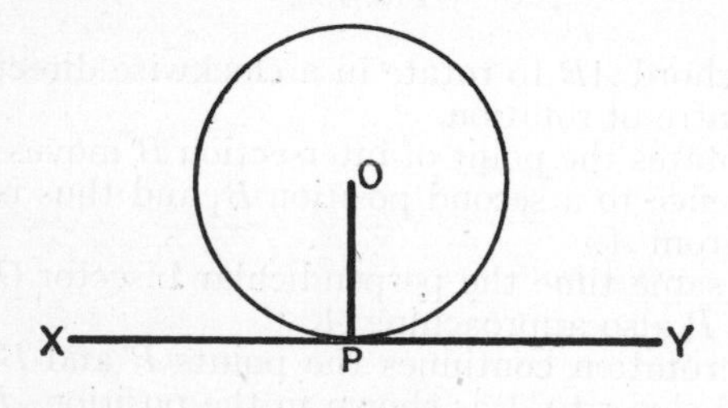

Fig. 137.

Join OP.

At P draw a straight line XY, perpendicular to OP.

Then by the conclusions of § 137 the straight line XY is a tangent to the circle.

139. Theorem. The tangents at the extremities of a chord of a circle are equal.

In Fig. 138 PQ is a chord of the circle, centre O.

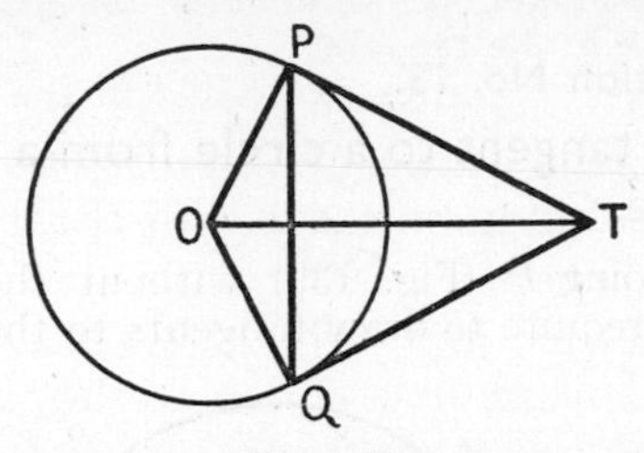

FIG. 138.

Join P and Q to O.

At P and Q draw perpendiculars to the radii OP and OQ.

These must intersect, as they are not parallel.

Let T be the point of intersection.

Then PT and QT are tangents at the extremities of the chord PQ.

Join OT.

In the **right-angled** $\triangle$**s OPT, OQT**:

(1) OT is a common hypotenuse;
(2) $OP = OQ$.

$\therefore$ the $\triangle$**s are congruent.** (§ 101)

In particular $TP = TQ$

also $\angle OTP = \angle OTQ$

and $\angle TOP = \angle TOQ$.

Hence, from a point outside a circle:

(1) *Two equal tangents can be drawn to the circle.*

(2) *The angle between the tangents is bisected by the straight line which joins their point of intersection to the centre.*

(3) *This straight line also bisects at right angles the chord which joins the points where they touch the circle.* (*The proof of this is left to the student.*)

Note.—PQ is called the **chord of contact** of the tangents TP, TQ.

It may further be noticed that since ∠s OPT, OQT are right angles they lie in semi-circles, of which OT is a common diameter. Hence the points O, P, T, Q are cyclic.

This fact enables us to perform the following important construction.

140. Construction No. 13.

To draw a tangent to a circle from a point without the circle.

From the point P (Fig. 139) without the circle ABC (centre O), we require to draw tangents to the circle.

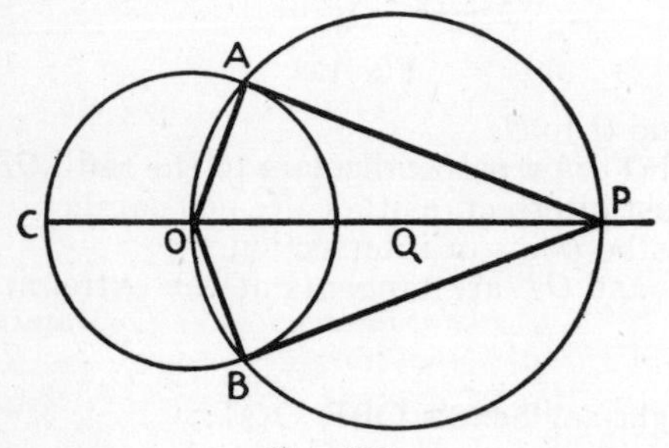

Fig. 139.

Construction.

Join P to O.

Bisect OP at Q.

On OP as diameter construct a circle $OAPB$, centre Q cutting the given circle at A and B.

Join PA, PB. These are the required tangents, there being two solutions to the problem.

Proof. Since OAP, OBP are semicircles.

∴ ∠s OAP, OBP are right angles (§ 135).

∴ **PA and PB are tangents to the circle, ABC.**

141. Circles which Touch One Another.

Two circles touch one another when they meet at a point, but their circumferences do not intersect.

There are two possible cases.

(1) **Of external contact**, as Fig. 140 (*a*), the circles being outside one another;

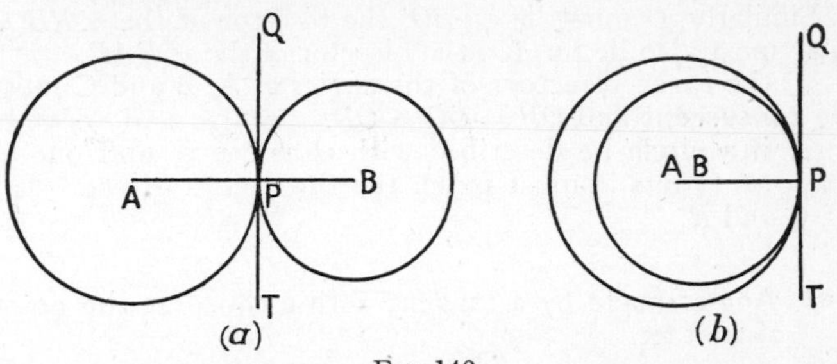

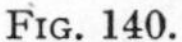
Fig. 140.

(2) **Of internal contact**, as Fig. 140 (*b*), one circle being inside the other.

Two facts are evident.

(*a*) In each case the circles have a **common tangent**, *QPT*, at the point of contact.

(*b*) The **line of centres**, *AB*, or *AB* produced, must pass through the point of contact, since a tangent is perpendicular to the radius at the point of contact.

142. Construction No. 14.

To inscribe a circle in a triangle.

An inscribed circle of a triangle, as of any rectilineal figure is **a circle to which the sides are tangential**, *i.e.*, it touches all sides.

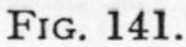
Fig. 141.

If *PQR* be the inscribed circle of the $\triangle ABC$ (Fig. 141) the method of obtaining the centre O, and the radius, can be deduced from §§ 139 and 140. CP and CQ are tangents from an external point C, and $\therefore OP = OQ$.

$\therefore O$ must lie on the bisector of the $\angle PCQ$ (§ 139).

Similarly, O must lie on BO, the bisector of the $\angle RBQ$.

It must also lie on AO, the bisector of the $\angle RAP$.

$\therefore$ **the three bisectors of the angles at A, B and C must be concurrent** and $OP = OQ = OR$.

$\therefore$ if a circle be described with O as centre and one of these as radius it must touch the three sides of the $\triangle$ at P, Q and R.

143. Angles made by a tangent with a chord at the point of contact.

PQ is a tangent to the circle ABC (Fig. 142). From the point of contact A, a chord AB is drawn dividing the circle into two segments, ABC (major segment) and ABD (minor segment).

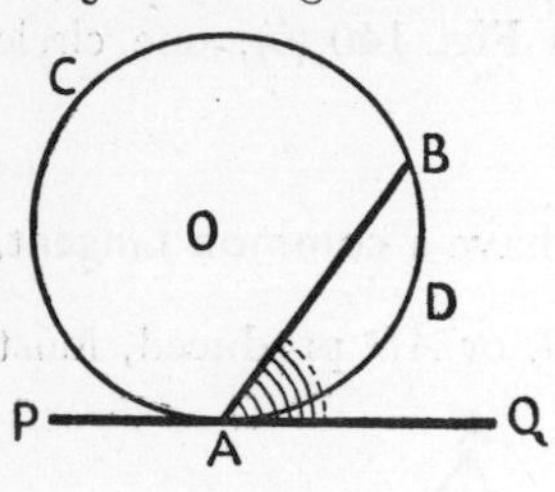

Fig. 142.

This chord at the point of contact makes two angles with the tangent, BAQ and BAP.

When considering the $\angle BAQ$, the segment which lies **on the other side of the chord AB**, *i.e.*, the major segment ACB, is called the **alternate segment** corresponding to this angle.

Similarly, if we are considering the $\angle BAP$ the alternate segment corresponding to it is the minor, *i.e.*, the segment BDA.

The following theorem shows an important connecting link between either angle and its alternate segment.

144. Theorem. The angles made by a chord of a circle with the tangent at an extremity of it are equal to the angles in the alternate segments.

In Fig. 143 PQ is a tangent to the circle $ACBD$ and AB is a chord drawn from the point of contact.

Draw the diameter AOC.

Join BC.

Then ∠ACB is an angle in the alternate segment ACB, corresponding to ∠BAQ.

Required to prove :

(1) $\angle BAQ = \angle ACB$.

Note.—It must be remembered that $\angle ACB$ is equal to any other angle which may be drawn in the segment ACB (Theorem, § 134). ∴ What is proved for $\angle ACB$ is also true for any other angle in the segment.

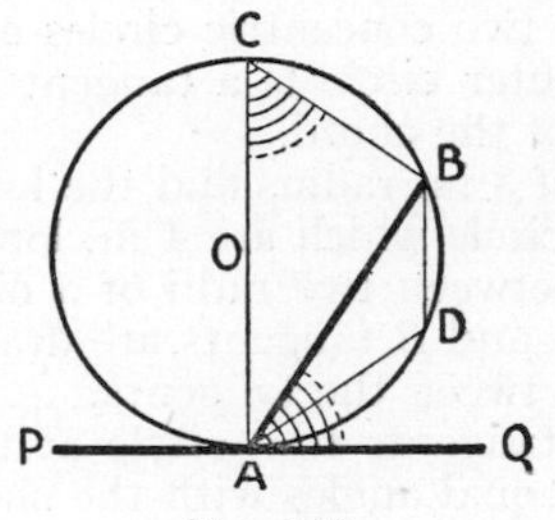

FIG. 143.

Proof. $\angle BAQ + \angle BAC$ = a right angle

also $\angle BCA + \angle BAC$ = a right angle

(since $\angle ABC$, being the angle in a semi-circle, is a right angle).

Subtracting $\angle BAC$ from each:

$$\therefore \quad \angle BAQ = \angle BCA.$$

(2) Let the point D be taken in the minor segment.

$\angle BDA$ is the angle in the corresponding alternate segment to $\angle PAB$.

∴ it is required to prove ∠PAB = ∠BDA.

Now $\angle PAB + \angle BAQ$ = two right angles

and $\angle BDA + \angle BCA$ = two right angles. (§ 136)

But $\angle BAQ$ was proved equal to $\angle BCA$ (1st part).

$$\therefore \quad \angle PAB = \angle BDA.$$

Exercise 16

1. PQ is a straight line which lies without the circle ABC. Show how to draw a tangent to the circle which is parallel to PQ. How many such tangents can be drawn?

2. In a circle of radius 1·5 in. construct a triangle with all its vertices on the circumference and having two of its angles 50° and 70°.

3. The radius of a circle is 1·5 in. From a point 2·5 in. from the centre a tangent is drawn to the circle. Find the length of this tangent.

4. The radii of two concentric circles are 3 in. and 4 in. A chord of the outer circle is a tangent to the inner one. Find the length of the chord.

5. In a circle of 3 in. radius find the locus of the centres of chords of the circle which are 4 in. long.

6. The angle between two radii of a circle, OA and OB, is 100°. From A and B tangents are drawn meeting at T. Find the angle between the tangents.

7. Prove that tangents to a circle at the extremities of any chord make equal angles with the chord.

8. Two circles are concentric. Prove that the tangents drawn to the inner circle from any point on the circumference of the outer circle are equal in length.

9. Find the locus of the ends of tangents of the same length which are drawn to a fixed circle.

10. Find the locus of the centre of circles which touch a fixed straight line at a given point.

11. Find the locus of the centres of circles which touch two intersecting straight lines.

12. Construct a triangle with sides 1·5 in., 2 in. and 3 in. in length. Then draw the inscribed circle.

CHAPTER 20

SIMILAR FIGURES. RATIO IN GEOMETRY

145. Similar triangles.

When the conditions under which triangles are congruent were examined (§ 47) it was pointed out that triangles with all three corresponding angles equal were not necessarily congruent. For this to be the case at least one pair of corresponding sides must also be equal.

In Fig. 144 are three triangles with corresponding angles

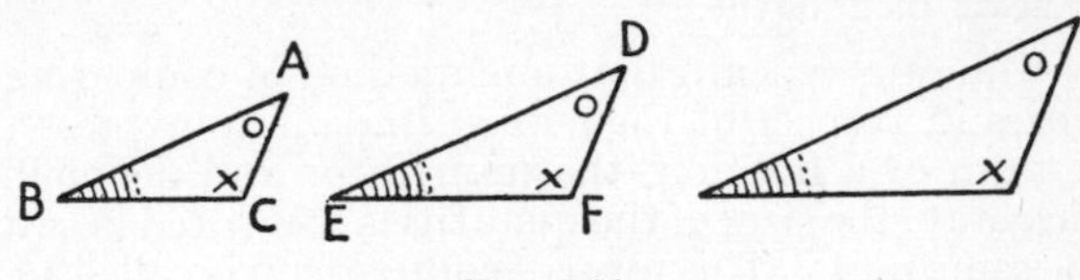

FIG. 144.

equal. The three triangles are of different sizes, but of the **same shape.** They are copies, one of another, on different scales.

Such triangles are called similar triangles.

In Fig. 145 is indicated a method by which a number of similar triangles can readily be drawn.

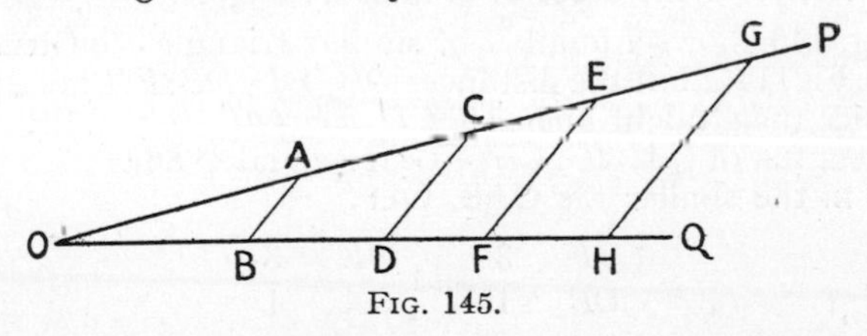

FIG. 145.

POQ is any angle and *OAB* a triangle formed by drawing any straight line *AB* to meet the two arms. From other points on *OQ*, such as *D*, *F*, *H* . . . draw straight lines *DC*,

FE, HG . . . parallel to *AB* and intersecting *OP* as shown, thus forming $\triangle$s *ODC, OFE, OHG*. . . .

The parallel straight lines *AB, CD, EF, GH* . . . being cut by the transversal *OQ* the corresponding angles at *B, D, F, H* . . . are equal.

Similarly, the angles at *A, C, E, G* . . . are equal.

$\therefore$ $\triangle$s *OAB, OCD, OEF, OGH* . . . have **corresponding angles equal.**

$\therefore$ **they are similar triangles.**

Triangles which thus have corresponding angles equal are said to be equiangular to one another.

Hence, the definition of similar triangles may be stated thus: *Triangles which are equiangular to each other are called similar triangles.*

146. Ratios of Lengths.

In arithmetic we learn that one method of comparing two quantities in respect of their magnitude is to express them in the form of a fraction, the numerator and denominator of which state the sizes of the quantities measured in suitable and the **same** units. This form of comparison is called a **ratio.**

Hence, when we speak of the **ratio of two straight lines** we mean the ratio of the numbers which express the measures of their lengths in terms of the same unit. Similarly, by the **ratio of the areas of two triangles** we mean the ratio of the numbers which express these areas in the same square units.

147. Ratios of the Sides of Similar Triangles.

Fig. 146 shows a number of similar triangles constructed as in Fig. 145, but the distances *OB, BD, DF, FH* are equal.

Since the straight lines *AB, CD, EF, GH* are parallel, then the lengths of *OA, AC, CE, EG* are equal (§ 86).

$\therefore$ **in the similar $\triangle$s OAB, OEF.**

$$\frac{OF}{OB} = \frac{3}{1} \text{ and } \frac{OE}{OA} = \frac{3}{1}.$$

$$\therefore \quad \frac{\mathrm{OF}}{\mathrm{OB}} = \frac{\mathrm{OE}}{\mathrm{OA}},$$

i.e., these sides are proportional.

Again, in the similar △s OGH, OEF.

$$\frac{OH}{OF} = \frac{4}{3} \text{ and } \frac{OG}{OE} = \frac{4}{3}.$$

$$\therefore \quad \frac{\mathrm{OH}}{\mathrm{OF}} = \frac{\mathrm{OG}}{\mathrm{OE}}.$$

Also, by drawing straight lines *EK*, *CL*, *AM* parallel to *OQ* it may be shown that:

$$\frac{EF}{AB} = \frac{OF}{OB} = \frac{3}{1}$$

and

$$\frac{GH}{EF} = \frac{OH}{OF} = \frac{4}{3}.$$

Similar conclusions may be reached with respect to other

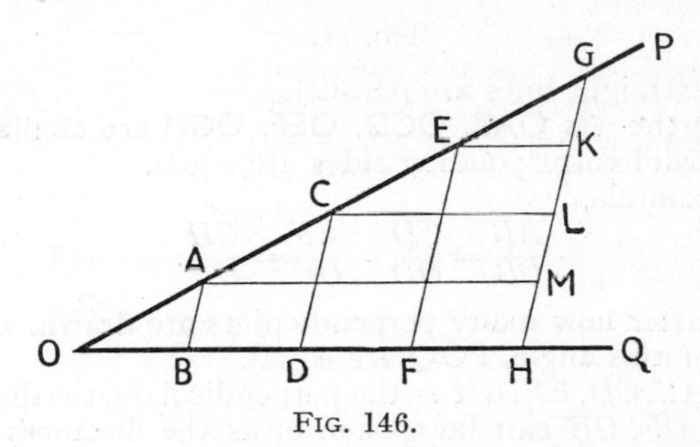

FIG. 146.

pairs of the four triangles in the figures. Hence, it is reasonable to conclude from all the above and similar results that:

The corresponding sides of similar triangles are in the same ratio, *i.e.*, the sides are proportional.

For example, in the similar △s *ABC*, *DEF* in Fig. 144,

$$\frac{AB}{DE} = \frac{BC}{EF} = \frac{AC}{DF}.$$

148. Fixed Ratios Connected with Angles.

(1) **The tangent.** There is a special case of the above conclusions which is of very great practical importance.

Draw **any** angle as POQ (Fig. 147).

Take a series of points A, C, E, G on one arm OP and draw AB, CD, EF, GH perpendicular to the other arm OQ.

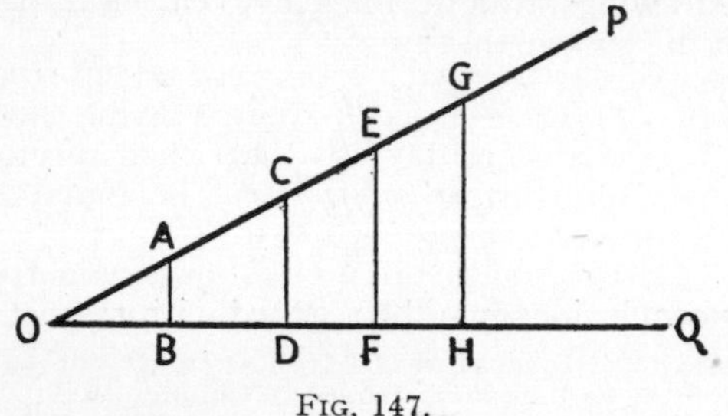

FIG. 147.

These straight lines are parallel.

Hence **the $\triangle$s OAB, OCD, OEF, OGH are similar,** and the ratios of corresponding sides are equal.

For example,

$$\frac{AB}{OB} = \frac{CD}{OD} = \frac{EF}{OF} = \frac{GH}{OH}.$$

No matter how many perpendiculars are drawn, **all such ratios for this angle, POQ** are equal.

With AB, CD, EF, GH as the perpendiculars, the distances OB, OD, OF, OH can be spoken of as **the distances intercepted on the arm OQ.**

Hence, for all such cases it is true to say that **the ratio**

$$\frac{\text{perpendicular drawn from one arm}}{\text{distance intercepted on the other arm}}$$

is constant for the angle POQ.

The angle chosen was **any** angle, consequently a similar conclusion can be reached for any other angle, *i.e.*,

For any angle the ratio of the perpendicular drawn from any point on one arm to the distance,

from the vertex, intercepted on the other arm, is constant for the angle.

This constant ratio is called the tangent of the angle.
Tangent is usually abbreviated to " tan ".
Thus, in Fig. 147,

$$\tan POQ = \frac{AB}{OB} = \frac{CD}{OD} = \frac{EF}{OF}, \text{ etc.}$$

Every angle has its own particular tangent and can be identified by it. Tables are constructed giving the tangents of angles between 0 and 90°, so that when the tangent is known, the angle corresponding may be found from the tables, and vice-versa.

For the further treatment of this Trigonometry should be consulted.

Note.—The term tangent as used above must not be confused with the tangent to a circle as defined in Chapter 19.

(2) **The sine and cosine.**

Two other constant ratios connected with an angle are given by taking the **ratios of each of the sides, in turn, containing the right angle, to the hypotenuse.**

In Fig. 147 the ratio of **the side opposite the angle** to the **hypotenuse** is the same for each of the triangles formed. Clearly

$$\frac{AB}{OA} = \frac{CD}{OC} = \frac{EF}{OE} = \frac{GH}{OG} = \frac{\text{side opposite}}{\text{hypotenuse}}.$$

This ratio is called the **sine** of the angle (abbreviated to " sin ").

Thus $$\sin POQ = \frac{AB}{OA} = \frac{CD}{OC} = \frac{EF}{OF}, \text{ etc.}$$

Also, the **ratio of the side adjacent to the angle to the hypotenuse** is the same for each of the triangles.

Thus

$$\frac{OB}{OA} = \frac{OD}{OC} = \frac{OF}{OE} = \frac{OH}{OG} = \frac{\text{side adjacent}}{\text{hypotenuse}}.$$

This constant ratio is called the **cosine** of the angle, abbreviated to " cos ".

Thus $$\cos POQ = \frac{OB}{OA} = \frac{OD}{OC}, \text{ etc.}$$

It will be noticed that since the hypotenuse is the greatest side of the triangle, **both sine and cosine must be numerically less than unity.** The tangent, however, may have any value.

149. Other Similar Figures.

The term "similar", in the sense it is used above, is not confined to triangles. All rectilineal figures, with the same number of sides, may be similar, provided that they conform to the necessary conditions stated above for triangles, viz.:

(*a*) All corresponding angles must be equal.
(*b*) Corresponding sides must be proportional.

In the case of triangles, if (*a*) is true, (*b*) must follow, as we have seen, so that it is sufficient to know that triangles are equiangular to one another. But with other rectilineal figures both conditions must be satisfied, before it can be

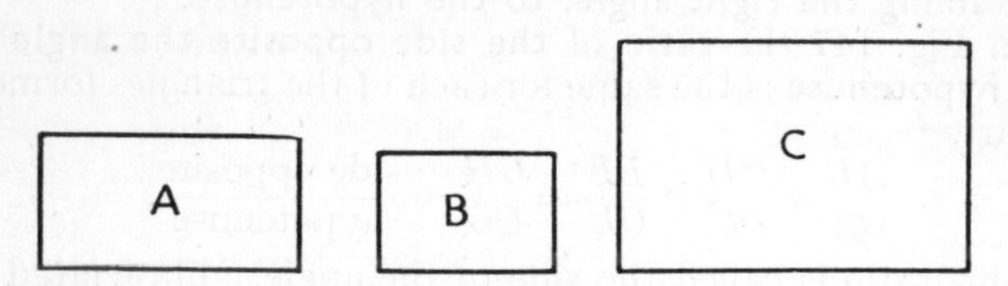

Fig. 148.

said that they are similar. If, however, they are equilateral, as with equilateral triangles, the ratios of corresponding sides are the same and corresponding angles must be equal.

Thus **all squares are similar, but rectangles, though equiangular, are not similar unless the ratios of corresponding sides are also equal.** Thus, in Fig. 148, the rectangles *A* and *B* are not similar, since the ratios of corresponding sides are obviously not equal. But *B* and *C* are similar, for the sides of B will be found to be one half the corresponding sides of C.

Regular polygons, such as hexagons, pentagons, etc., are similar, but polygons which are not regular may be similar only if conditions (*a*) and (*b*) are satisfied.

Generally when two figures are similar their "**shapes**" are the same: one is a copy of the other on a different scale.

All drawings and models, when not full size, are **drawn or constructed to scale.** When thus drawn or constructed they are similar. Angles are copied exactly and the ratio of corresponding distances is that of the scale employed. If, for example, a model is made on a scale of an inch to a yard, lengths in the model will in all cases be $\frac{1}{36}$ of the corresponding length of the original.

Pictures appearing on the cinema screen are greatly enlarged copies of small photographs on the film, all parts being enlarged in the same ratio. The pictures are therefore similar.

The picture of the west front of a cathedral shown in the frontispiece is **similar** in every detail to a picture of the same building which is ten times its size. Both are similar in appearance to the building itself.

150. Construction No. 15.

To divide a straight line in a given ratio.

Example: Divide the straight line AB (*Fig.* 149) *in the ratio of* 3 : 2.

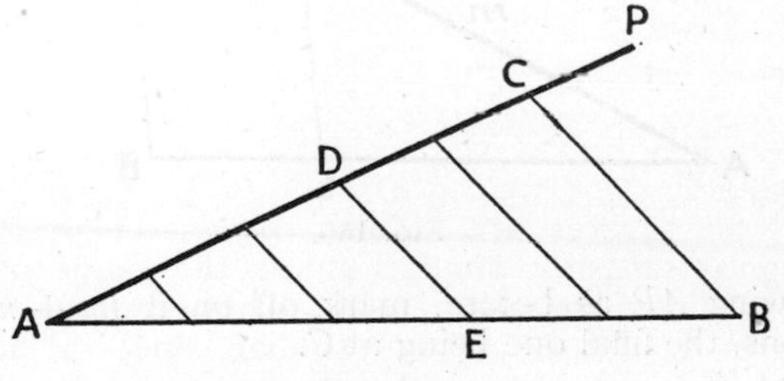

FIG. 149.

The solution of the problem depends directly upon Construction No. 9, § 87.

First, divide AB into $(3 + 2) = 5$ equal parts as follows.

From A draw AP at any convenient angle.

Along AP mark off with dividers 5 equal distances.

Join CB. From the points of division on AC draw straight lines parallel to CB to meet AB.

Then AB is divided into 5 equal parts.

Let DE be the straight line joining the 3rd points from A on AC and AB.

Let x be the length of each of the equal parts of AB.

Then $$AE = 3x$$
and $$EB = 2x.$$

$$\therefore \frac{AE}{EB} = \frac{3x}{2x} = \frac{3}{2}.$$

∴ the straight line AB is divided at E in the ratio of 3 : 2.

Note.—In practice it is necessary to draw CB and DE only.

The method may be generalised thus.

Let it be required to divide the straight line AB (Fig. 150) in the ratio m : n.

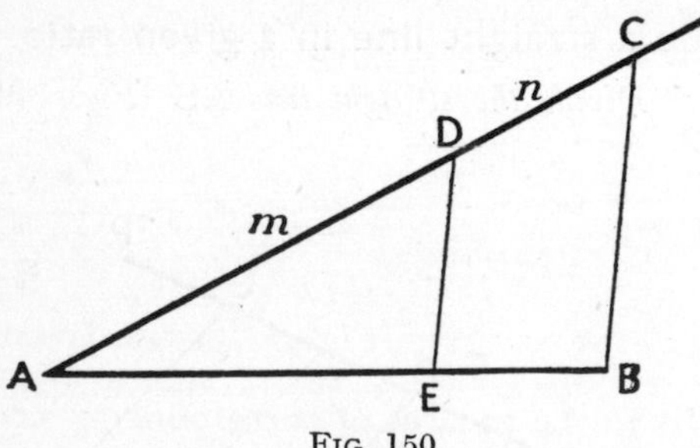

Fig. 150.

Drawing AP as before, mark off on it $(m + n)$ equal divisions, the final one being at C.

Join CB.

From the mth point of division on AC, viz., D, draw DE parallel to CB.

Then the ratio of $\dfrac{AD}{DC} = \dfrac{m}{n}$.

$\therefore$ the ratio of $\dfrac{AE}{EB} = \dfrac{m}{n}$.

Corollary. Since ACB is a triangle it may be concluded in general that, **if a straight line be drawn parallel to one of the sides of a triangle, it cuts the other two sides in the same ratio.** See Theorem in § 84, for a special case.

151. Areas of Similar Figures.

The areas of similar figures are proportional to the squares of corresponding sides.

The simplest example of this principle, and one with which the student is acquainted, is that of the **square.**

If the side of a square is doubled the area is increased 4 times.

If it be increased 3 times the area is increased 9 times.

These and similar examples can readily be seen by observation of Fig. 90.

In general, if the side of a square be increased n times the area is increased n^2 times.

Again, the formula for the area of a **circle,** viz., $A = \pi r^2$ indicates that the **area is proportional to the square of the radius.**

Thus, if the radii of two circles are r_1 and r_2.

Then $$\text{ratio of areas} = \frac{\pi r_1^{\,2}}{\pi r_2^{\,2}} = \frac{r_1^{\,2}}{r_2^{\,2}}.$$

Area of a triangle. In Part II a geometrical proof is given of the theorem that "**the areas of similar triangles are proportional to the squares of corresponding sides**" (see Theorem 64).

Exercise 17

1. In a $\triangle ABC$, PQ is drawn parallel to BC and cutting AB and AC in P and Q. $AB = 5$ in. and $AC = 8$ in. Also, the ratio of $AP : PB = 2 : 3$. Find the lengths of AP, PB, AQ, QC. If $PQ = 6$ in., find BC.

2. OA, OB, OC are the bisectors of the angles A, B, C of a $\triangle ABC$, O being their point of intersection. From a point P on AO, PQ is drawn parallel to AB, meeting BO in Q. From Q, QR is drawn parallel to BC, meeting OC in R. Join PR. Prove that the $\triangle$s ABC and PQR are similar.

3. In a circle two chords AB and CD intersect at O. AD and CB are drawn. Prove $\frac{AO}{OB} = \frac{DO}{OC}$. (*Hint.*—Join DB.)

4. Divide a straight line 4 in. long in the ratio 4 : 3.

5. Trisect a line which is 5 in. long. If the perimeter of an equilateral triangle is 5 in., construct the triangle.

6. The perimeter of a triangle is 7 in. and its sides are in the ratio of 3 : 4 : 5. Construct the triangle.

7. Two $\triangle$s ABC, DEF are similar and the altitudes from A and D are 3 in. and 4 in. respectively. If the area of the smaller triangle is 22·5 sq. in., find the area of the larger triangle.

8. The area of one square is twice that of another. Find the ratio of their sides.

9. Equilateral triangles are described on the side and diagonal of a square. Find the ratio of their areas.

10. Construct an angle of 50°. From three points on one arm draw perpendiculars to the other arm. Hence, find by measurement in each case the tangent of 50° and find the average of the three results.

11. Using the results of §§ 101 and 102 find the sine, cosine and tangent of (1) 30°, (2) 60°, (3) 45°.

CHAPTER 21

RELATIONS BETWEEN THE SIDES OF A TRIANGLE

(This chapter might be omitted by beginners)

152. Extensions of the Theorem of Pythagoras.

In Chapter 13 the very important law, known as the **Theorem of Pythagoras**, was established, in which is stated the relations which exist between the sides of a right-angled triangle. We now extend the investigations to ascertain what similar laws connect the sides of triangles which are not right-angled, *i.e.*, they are obtuse-angled or acute-angled triangles.

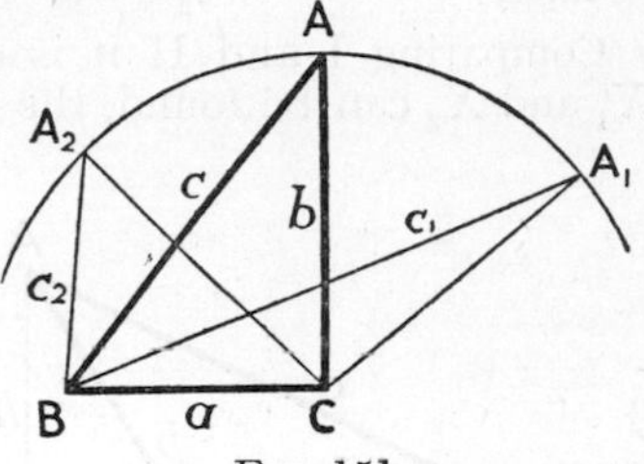

FIG. 151.

In Fig. 151 ABC is a right-angled triangle, C being the right angle.

Denoting the sides in the usual way by a, b, c, then, by the *Theorem of Pythagoras*,

$$c^2 = a^2 + b^2.$$

(1) Obtuse-angled triangles.

With C as centre and CA as radius, describe an arc of a circle. On the same side as the right angle take a point A_1 and join to B and C.

The $\angle A_1CB$ is clearly obtuse and the $\triangle A_1BC$ obtuse-angled. Comparing the sides of this triangle with those of the $\triangle ABC$ it is seen that

$$A_1C = AC.$$

BC is common to each, but A_1B is greater than AB.

Denoting A_1B by c_1, then c_1 must be greater than c.

$$\therefore \quad c_1^2 > (a^2 + b^2).$$

Let X_1 represent the amount by which c_1^2 exceeds $(a^2 + b^2)$.

Then $$c_1^2 = a^2 + b^2 + X_1 \quad . \quad . \quad . \quad . \quad (I)$$

(2) Acute-angled Triangles.

On the arc previously drawn, but on the other side of AC_1 take a point A_2, so that $\angle A_2CB$ is an acute angle and the $\triangle A_2BC$ is an acute-angled triangle.

Denoting A_2B by c_2 it is clear that c_2 is less than c.

$$\therefore \quad c_2^2 < (a^2 + b^2).$$

Let X_2 represent the amount by which it is less.

Then $$c_2^2 = a^2 + b^2 - X_2 \quad . \quad . \quad . \quad (II)$$

Comparing I and II it is evident that if the values of X_1 and X_2 can be found, the relations between the sides of the obtuse-angled and acute-angled triangles can be definitely established.

In obtaining these values, use will be made of some of the methods and results of elementary algebra.

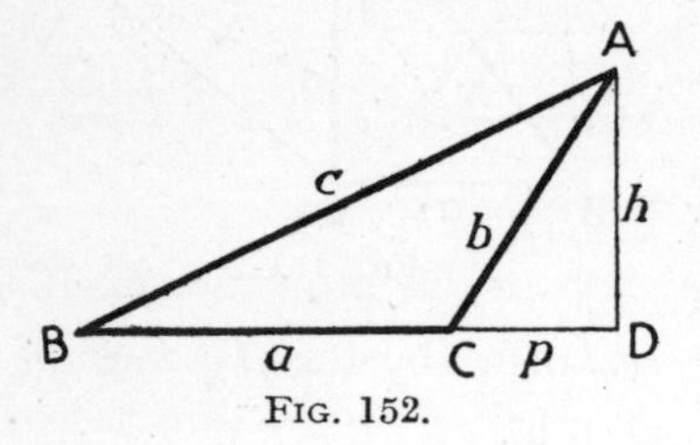

Fig. 152.

(3) The Obtuse-angled Triangle.

To simplify the method an obtuse-angled triangle is drawn separately, as in Fig. 152.

From A draw AD perpendicular to BC produced.

Then AD is the altitude or height of the triangle, when A is a vertex.

Then CD is the projection of AC on BC produced (*see* § 30).

Let $$AD = h \quad \text{and} \quad CD = p.$$
Then $$BD = (a + p).$$

Applying the Theorem of Pythagoras to the $\triangle ABD$,

$$c^2 = (a + p)^2 + h^2.$$

But by Algebra $(a + p)^2 = a^2 + 2ap + p^2$ (*see* Appendix).

$$\therefore \quad c^2 = a^2 + 2ap + (p^2 + h^2).$$

But in $\triangle ACD$ $\quad b^2 = p^2 + h^2.$

Substituting this value for $p^2 + h^2$ in the result above.

Then $\quad c^2 = a^2 + 2ap + b^2 \quad . \quad . \quad . \quad \text{(A)}$

Thus, we find, comparing with (I), that

$$X_1 = 2ap.$$

(*b*) **Acute-angled triangle.**

In the acute-angled $\triangle ABC$ (Fig. 153), AD is the per-

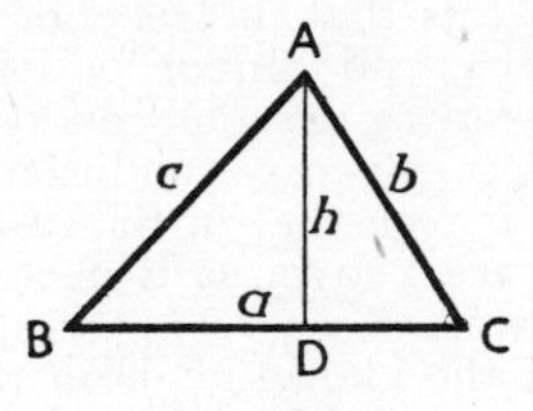

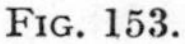
FIG. 153.

pendicular from A on BC and DC is the projection of AC on BC.

Using the same letters as before,

Let $\quad h =$ altitude or height

$\quad p =$ projection of AC, *i.e.* DC.

Then $\quad BD = a - p.$

From $\triangle ABD$, $\quad h^2 = c^2 - (a - p)^2.$ (§ 101)

From $\triangle ACD$, $\quad h^2 = b^2 - p^2.$ (§ 101)

$$\therefore \quad c^2 - (a - p)^2 = b^2 - p^2$$

and $\quad c^2 = (a - p)^2 + b^2 - p^2.$

From Algebra

$$(a - p)^2 = a^2 - 2ap + p^2 \quad \text{(see Appendix).}$$

$\therefore$ Substituting

$$c^2 = (a^2 - 2ap + p^2) + b^2 - p^2$$

and $\quad c^2 = a^2 + b^2 - 2ap \quad . \quad . \quad . \quad . \quad . \quad \text{(B)}$

$\therefore$ from (II),

$$X_2 = 2ap.$$

Summarising the results (A) and (B).

(1) For obtuse-angled triangles

$$c^2 = a^2 + b^2 + 2ap \quad . \quad . \quad . \quad \text{(A)}$$

(2) For acute-angled triangles

$$c^2 = a^2 + b^2 - 2ap \quad . \quad . \quad . \quad \text{(B)}$$

Thus (A) and (B) differ only in the sign of the term $2ap$.

153. Use of the Formulæ in Calculations.

In practice the difficulty in using the above formula for the evaluation of c is that, in general, neither the value of h or p is known, and cannot be determined without further data. Referring to the conditions under which triangles are fixed, in this case condition (A) for congruent triangles (§ 47), it is seen that in the case under consideration, **if two sides are known, it is necessary to know the included angle as well.**

To make use of the known angle in the above formula, we refer back to § 148, in which it was shown that in a right-angled triangle there are ratios between the sides which are constant for any given angle. Thus if the angle is known, we can obtain from tables the values of its sine, cosine or tangent. In the above case (Fig. 153) it is seen that

$$\frac{CD}{AC} \text{ i.e. } \frac{p}{b} = \cos ACB,$$

whence

$$p = b \cos C.$$

Consequently, in formulæ (A) and (B) above, p can be replaced by $b \cos C$.

Thus formula (B) becomes

$$c^2 = a^2 + b^2 - 2ab \cos C \quad . \quad . \quad . \quad \text{(B)}$$

It is proved in Trigonometry that **the cosine of an obtuse angle is equal to — (cosine of its supplement).**

i.e. in Fig. 152 $\cos ACB = - \cos ACD$

Consequently, in formula (A) on substituting

$$p = - b \cos C,$$

the formula becomes

$$c^2 = a^2 + b^2 - 2ab \cos C \quad . \quad . \quad . \qquad (A)$$

Thus, in both the cases of obtuse- and acute-angled triangles the formula is

$$c^2 = a^2 + b^2 - 2ab \cos C.$$

154. Example.

The following example illustrates one of the many ways in which the above formula may be employed practically.

In Fig 154 (not drawn to scale) A, C and B represent the positions of three towns on the shore of a harbour. The distance of A from C is 5 miles and of B from C 4 miles. The angle ACB is 75°. What is the distance of B from A?

(Given $\cos 75° = 0{\cdot}2588$.)

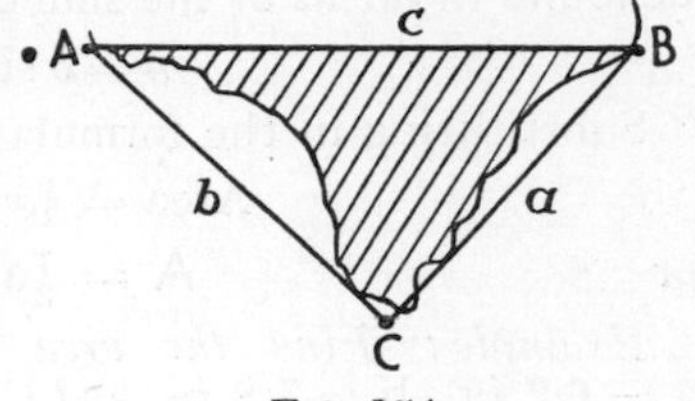

FIG. 154.

In practice, distances on land such as AC and BC, are readily determined by surveying methods and the angle ACB is found by a theodolite. When these are known a distance such as AB which is inaccessible for direct measurement can be determined by using the above formula.

Using the same notation for the sides of the $\triangle ABC$ as in Fig. 153, we substitute in the formula

$$c^2 = a^2 + b^2 - 2ab \cos C.$$

The values $a = 4$, $b = 5$, $\cos C = 0{\cdot}2588$.

Then
$$\begin{aligned} c^2 &= 4^2 + 5^2 - 2 \times 4 \times 5 \times 0{\cdot}2588 \\ &= 16 + 25 - 40 \times 0{\cdot}2588 \\ &= 30{\cdot}648 \end{aligned}$$

and $c = 5{\cdot}53$ miles (approx.).

For many developments of this useful formula a book on *Trigonometry* must be studied.

155. Area of a Triangle.

Referring to Fig. 153 it will be seen that

$$\frac{h}{b} = \sin ACB.$$

$$\therefore \quad \text{h} = \text{b} \sin \text{ACB}.$$

Thus, a way is found for finding the altitude of a triangle in terms of the given sides and the sine of the included angle.

In § 96 it was found that the area of a triangle can be expressed in the form

$$\text{Area} = \tfrac{1}{2}bh.$$

But, again, h is not always known. If, however, two sides and the included angle are given the **value of h can be found in terms of the sine of the angle**, viz.

$$h = b \sin C.$$

Substituting in the formula for the area,

$$\text{Area} = \tfrac{1}{2}a \times b \sin C,$$

or

$$\text{A} = \tfrac{1}{2}\text{ab} \sin \text{C}.$$

Example: Find the area of the triangle ABC *when* a = 6·2 *in.*, b = 7·8 *in. and* C = 52° (sin 52° = 0·7880).

Since $A = \frac{1}{2}ab \sin C$,

substituting $A = \frac{1}{2} \times 6{\cdot}2 \times 7{\cdot}8 \times 0{\cdot}7880$

$= 19{\cdot}1$ sq. in. (approx.).

Exercise 18

1. Write down a formula, similar to that in § 153, for finding the third side of a triangle when the following are known :

   (a) $b, c, \angle A$. (b) $a, c, \angle B$.

2. Write down a formula for the area of a triangle similar to that of § 155 when the elements of the triangle which are given are as in the previous question.

3. In an acute-angled triangle ABC, find the side c, when $a = 10$, $b = 11$, $\cos C = 0{\cdot}3501$.

4. In an obtuse-angled triangle ABC, find c when $a = 2$ in., $b = 3$ in., $\cos C = -0{\cdot}2501$.

5. The sides of a triangle are $a = 8$ in., $b = 9$ in., $c = 12$ in. Find $\cos C$.

6. Find a when $b = 19$, $c = 26$ and $\cos A = 0{\cdot}4662$.

7. Find the area of a triangle in which $a = 39$ in., $b = 53$ in. and $\sin C = 0{\cdot}8387$.

8. Find the area of a triangle ABC when $AB = 14$ in., $BC = 11$ in. and $\sin B = 0{\cdot}9397$.

CHAPTER 22

SYMMETRY IN GEOMETRY

156. The Meaning of Symmetry.

What is called "symmetry" is an essential factor in most forms of pictorial design. The architect, in designing the façade of a building, the cabinet-maker in designing a cabinet or a chair, the potter planning a vase, all make use of symmetry, under suitable conditions, in some form or another. By this is meant generally that if a straight line be drawn down the middle of the design, in which symmetry is an essential factor, the two parts into which the design is divided are alike. Any particular form or shape on one side of the middle line is balanced by the same feature on the other side.

This may be observed in the picture of the beautiful west front of the cathedral at Florence, which forms the frontispiece to this book. If an imaginary straight line be drawn vertically from the highest point of the centre of the building (not the tower) where a cross is erected, it will be seen that the two parts on either side of it are identical. There is a perfect balancing of component features; every feature on one side is balanced by a similar feature on the other. For example, a circular window on the right-hand side of the centre-line, or axis, is balanced by a similar window on the other side, of the same size and at the same distance from the axis.

Symmetry is a feature of mens' faces, or of most of them. If an imaginary line be drawn through the centre of the forehead and down the centre of the bridge of the nose the two parts of the face on either side are usually identical.

157. Symmetry in Geometrical Figures.

The above examples of symmetry relate to solid objects, but a more precise form of symmetry is to be seen in many geometric figures in a plane.

Examples are shown in Fig. 155.

(1) A **circle** is divided by any diameter into two parts which are similar. Thus it is said to be symmetrical about any diameter, such as AB (Fig. 155 (*a*)).

(2) Similarly an **isosceles triangle** (Fig. 155 (*b*)) is symmetrical about the straight line AD, which bisects

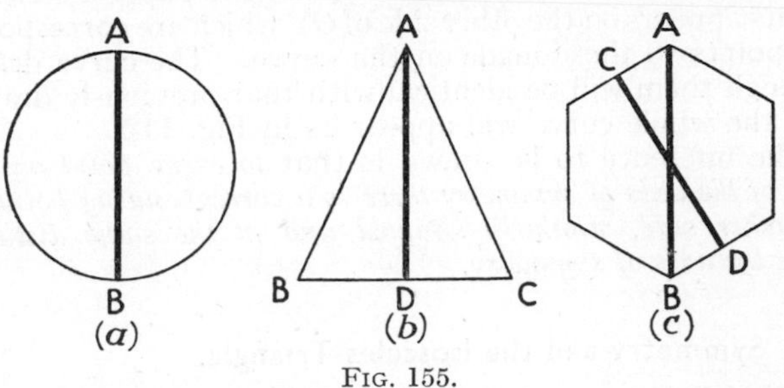

FIG. 155.

the vertical angle, and also bisects the base at right angles (see § 62).

(3) A **regular hexagon** (Fig. 155 (*c*)) is symmetrical about any diagonal, such as AB, or about a straight line such as CD which joins the mid-points of two opposite sides.

(4) An **ellipse** is symmetrical, as shown in Fig. 109, about the major axis AB, or the minor axis CD.

158. An Axis of Symmetry.

It has been stated above that there was symmetry about certain straight lines, *e.g.*, the circle about any diameter.

The straight line about which a figure is symmetrical is called an axis of symmetry.

A test which may be applied to many geometrical figures as to their symmetry is that, if they are folded about an axis of symmetry, the two parts of the figure coincide.

The parabola in Fig. 112 is a symmetrical curve. This is also evident from the method of drawing it, which is briefly referred to in § 116, and which is familiar to those who have studied the algebraical treatment of graphs.

The following experiment in folding will be found useful.

Draw the portion of the curve on the right-hand side of OY (Fig. 112), that is, for positive values of x. Then fold the paper exactly along OY. Now prick through a number of points on the curve. On opening out the paper a series of points appears on the other side of OY which are **corresponding points** to those made on the curve. The curve drawn through them will be identical with that previously drawn, and the whole curve will appear as in Fig. 112.

The inference to be drawn is that *to every point on one side of the axis of symmetry there is a corresponding point on the other side, similarly situated and at the same distance from the axis of symmetry.*

159. Symmetry and the Isosceles Triangle.

The following example in folding illustrates the use that may be made of symmetry in demonstrating the truth of certain geometrical theorems.

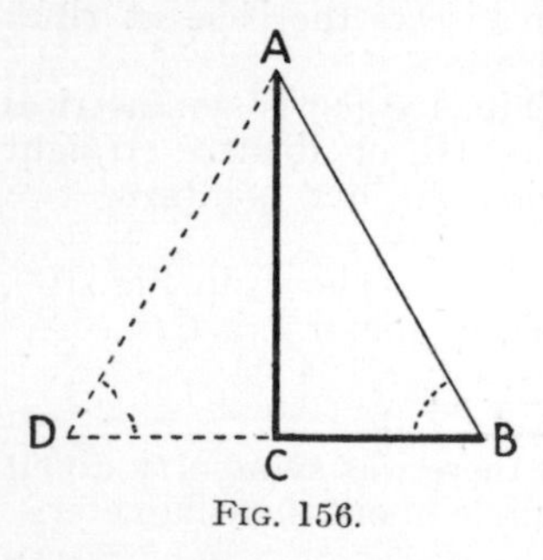

Fig. 156.

Construct a right-angled triangle such as ABC (Fig. 156).

Fold the paper carefully about AC.

Cut out the triangle and then open it out. The triangle ABD will appear, made up of the two right-angled $\triangle$s ACB, ACD.

Since the $\triangle$s ACB, ACD are identical, the $\triangle ABD$ is symmetrical about AC as an axis of symmetry (§ 157 (2)). Consequently,

(1) $AD = AB$ and the $\triangle ABD$ is isosceles.
(2) AC is the bisector of the vertical angle at A.
(3) $\angle ABC = \angle ADC$ (§ 62).
(4) AC is the perpendicular bisector of BD (*see* Theorem, § 62).

160. A knowledge of symmetry makes it possible for an architect or engineer, when preparing working drawings, to draw half of the figure only, on one side of the axis of symmetry. The other half, being identical with it, is frequently unnecessary, all the details being shown in the half which is drawn.

Exercise 19

1. Which of the following figures are symmetrical? What are the axes of symmetry? If there are more than one describe them :

(*a*) square.
(*b*) rectangle.
(*c*) two intersecting circles.
(*d*) a sector of a circle.
(*e*) trapezium.
(*f*) triangle with angles 45°, 45°, 90°.
(*g*) triangle with angles 30°, 60°, 90°
(*h*) a regular Pentagon.

2. Construct an irregular rectilineal figure, of five sides, which is symmetrical.

3. How many axes of symmetry are there in an equilateral triangle? State what they are.

4. Is a rhombus a symmetrical figure? If so, what is the axis of symmetry?

CHAPTER 23

PARALLEL PLANES

161. If the outside cover of a match-box be examined it will be noted that the top and bottom faces are two plane surfaces or planes which are always the same distance apart, and it is evident that they would never meet, no matter how far they might be extended. Thus, they satisfy a condition similar to that which must be fulfilled by straight lines which are parallel. The planes or surfaces are said to be **parallel.**

Another example nearer to hand is that of the two outside surfaces of the cover of this book, when it is closed and laid on the table. These are everywhere the same distance apart and will not meet if extended in any direction. They are parallel planes.

Parallel planes may thus be defined in the same way as parallel straight lines.

Definition. *Planes which do not meet when extended in any direction are called parallel planes.*

All horizontal planes are parallel (*see* § 28), but this is not necessarily true for vertical planes. The corner of a room marks the intersection of two vertical planes which meet; but the two opposite walls are generally parallel vertical planes.

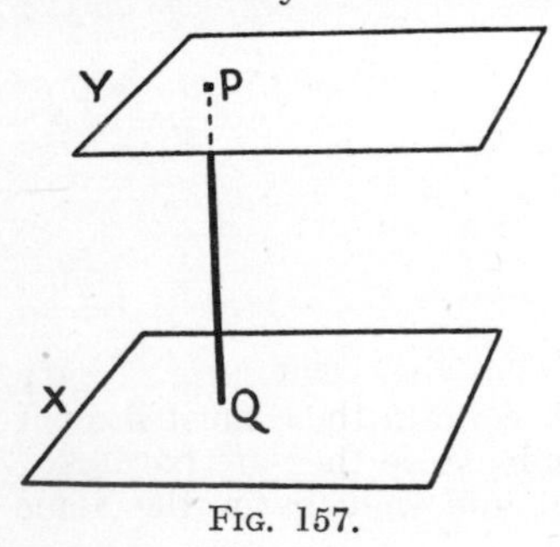

FIG. 157.

162. Planes to which the same straight line is perpendicular are parallel.

In Fig. 157 X and Y represent two plane surfaces or planes.

The straight line PQ is perpendicular to both planes (§ 29).

Then the planes are parallel.

No proof is offered for this statement, but it may reasonably be regarded as self-evident.

163. Theorem. If two parallel planes are cut by another plane, the lines of intersection with those planes are parallel.

Let X and Y (Fig. 158) be two parallel planes. These

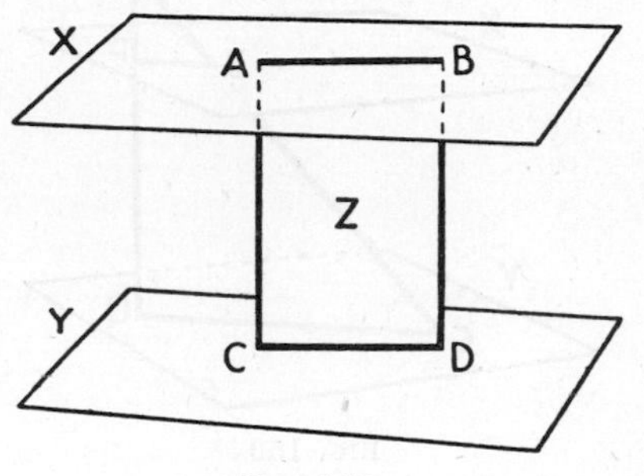

FIG. 158.

are cut by another plane Z which intersects X in AB and Y in CD.

Required to prove :

AB and CD are parallel.

Proof :

If AB and CD are not parallel they will meet if produced. Then the planes which contain them must meet if extended. But this is impossible, since they are parallel.

$\therefore$ AB and CD cannot meet and they lie in the same plane Z.

$\therefore$ AB and CD are parallel.

164. Theorem. If a straight line meets two parallel planes, it makes equal angles with them.

In Fig. 159 X and Y represent two parallel planes.

ABC is a straight line which meets them at B and C respectively.

From A draw APQ perpendicular to the two planes and meeting them in P and Q.

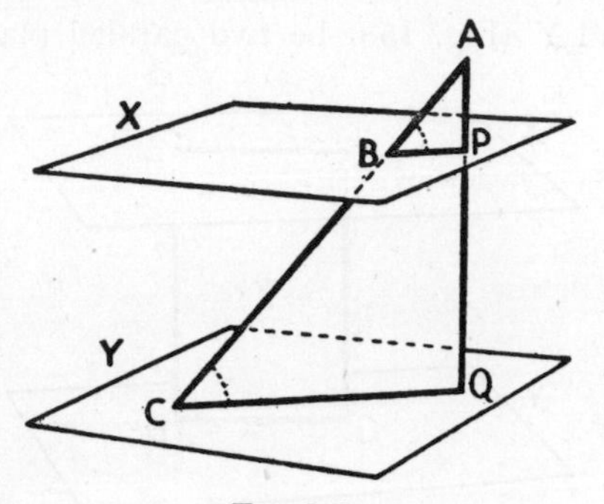

FIG. 159.

Join BP, CQ.

Then $\angle$s ABP, ACQ are the angles made by ABC with the two planes (§ 30).

Required to prove :

$$\angle ABP = \angle ACQ$$

Proof :

ACQ is a plane which meets the parallel planes X and Y in BP and CQ respectively.

$\therefore$ BP and CQ are parallel (§ 163).

They are cut by the transversal ABC.

$\therefore$ corresponding angles ABP and ACQ are equal, *i.e.*, the straight line ABC is equally inclined to both planes.

CHAPTER 24

PRISMS

165. Geometrical Solids.

With the exception of Chapters 4 and 23, the geometry of figures in one plane, *i.e.*, " **plane figures** " has, so far, been our main consideration. The next three chapters will be entirely concerned with the geometry of " **solids** ".

The meaning which is attached to the word " solid " in geometry was discussed in § 2, and the student should revise it now. The final statement of the paragraph is quoted as a summary.

" **A solid body, from the point of view of Geometry, is conceived as a portion of space enclosed and bounded by surfaces and the amount of space so occupied is called its volume.**"

The particular kinds of solids to be considered are what may be termed " *geometrical solids* ", *i.e.*, the surfaces which bound, or enclose them, are geometric figures, such as have been considered in previous chapters.

166. A cross-section of a Solid.

If the trunk of a tree be sawn through, the new surface which is thus exposed is called a " **cross-section** " of the trunk. In this case the " cross-section " will not be a regular geometric figure, such as a circle, but will be irregular in outline.

If however a regular solid, such as a stick of shaving soap, is cut through, the cross-section will be a regular figure. If a rectangular wooden block is sawn through the section will be a rectangle.

167. Prisms.

The rectangular block considered in § 2 is a common form of a geometrical solid. Fig. 160 represents a solid of this type.

Its boundary surfaces are six in number, consisting of three pairs of parallel rectangles. Each of these is perpendicular to the four planes which intersect it.

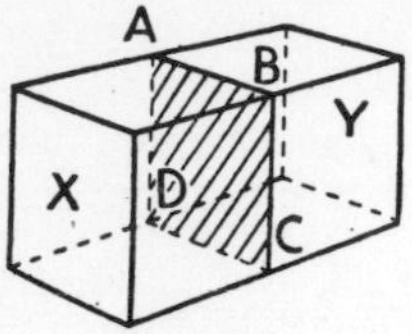

A **cross-section** is indicated by the shaded rectangle $ABCD$, which intersects four faces.

Fig. 160.

It is such that the **plane of the section is perpendicular to each of the four plane surfaces which it intersects.** It is therefore parallel to the pair of parallel end surfaces indicated by X and Y.

Such a section is called a normal section.

Its area will be the same as that of the parallel end surfaces. Any section which was not parallel would have a greater area.

When a normal cross-section is always of the same size and shape the solid is called a **prism.**

Consequently,

A prism is a solid of uniform normal cross-section.

When this cross-section is a rectangle, the solid is called a **rectangular prism**; when it is a square it is a **square prism.**

If all six faces are squares of equal size the solid is **a cube.**

When all the faces which meet intersect at right angles, the solid is called a **right prism.**

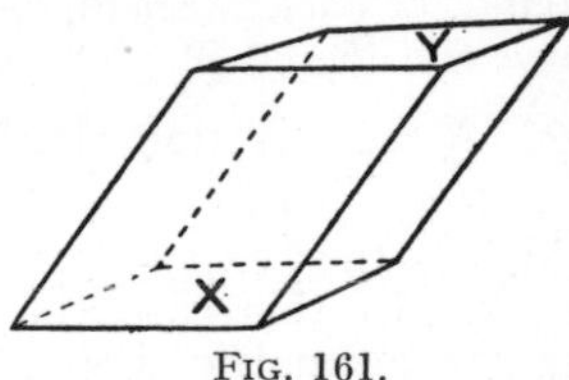

Fig. 161.

If the faces are not at right angles, as in Fig. 161, the solid is an **oblique prism.** Cross-sections parallel to the end faces such as X and Y will be similar figures to them.

Note.—The student may construct an oblique prism by squeezing the two opposite faces of the cover of a match box, as described in § 76.

168. Other Forms of Prisms.

The normal cross-section of a prism may be any regular figure: thus, we may have a **triangular prism,** in which the normal cross-section is a triangle. An example is shown in Fig. 162. Similarly, there may be hexagonal prisms, pentagonal prisms, octagonal prisms, etc.

169. The Cylinder or Circular Prism.

When the normal cross-section of a prism is a circle the solid is called a **cylinder.** Fig. 163 represents a cylinder and PQ a normal cross-section.

One end, such as ACB, may be regarded as a base. Its

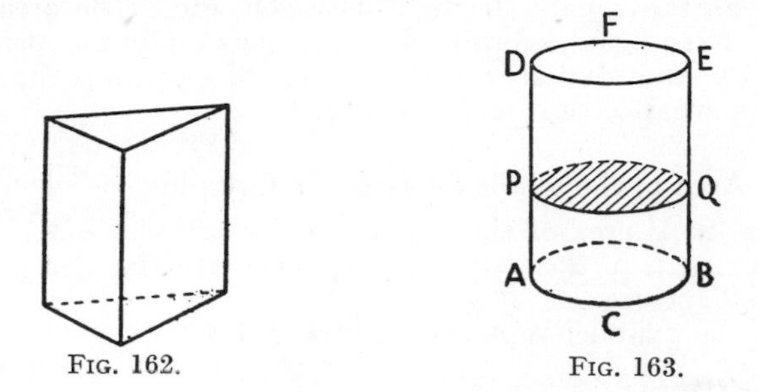

Fig. 162. Fig. 163.

plane is parallel to that of the other end DFE; all normal cross-sections are parallel to both.

Area of the curved surface.

A cylinder made of paper can be constructed as follows. Take a jam tin or jar, which is cylindrical, and wrap a piece of suitable paper exactly round it until the opposite edges of the paper meet. The paper then constitutes a cylinder. If it be unrolled the paper will be seen to be a rectangle, one side of which is the same in length as the height of the cylinder, and the other is equal to the circumference of the cylinder. It is thus evident that the

Area of this rectangle is the same as the area of the curved surface of the cylinder.

The cylinder in the form of the cylindrical pillar is prominent in the construction of many great buildings, such as cathedrals, churches, etc.

Fig. 164 is an illustration of a building in which cylindrical pillars and semi-circular arches combine to produce a beautiful and graceful effect. They also provide strength and endurance; this porch has existed for over 700 years.

170. Area of the whole Surface of a Prism.

With the exception of the cylinder, the faces and bases of prisms are rectilineal figures, the faces being a series of rectangles, and the ends regular geometrical figures, such as the square, triangle, hexagon, etc. The areas of these have been determined in previous chapters. Consequently, **the total area of the surface of a prism is the sum of the areas of the faces and ends.**

171. Area of the whole Surface of a Cylinder.

The total area of the surface of the cylinder is the sum of the areas of the two ends together with the area of the curved surface.

In the cylinder represented in Fig. 165

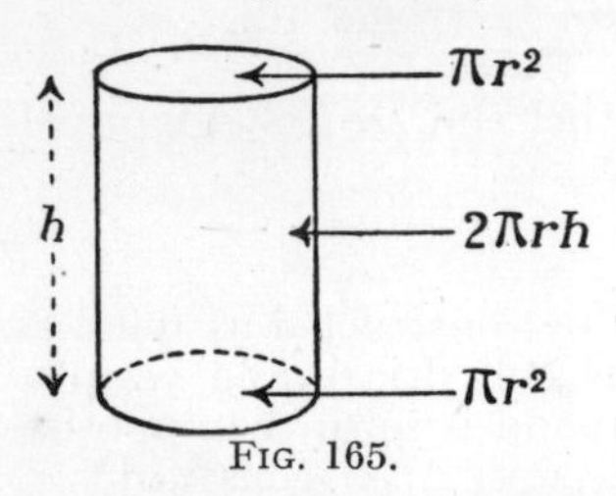

Fig. 165.

Let r = radius of normal cross-section.
h = height of cylinder.

Area of ends.

Area of each end $= \pi r^2$.

$\therefore$ area of two ends $= 2\pi r^2$.

Area of curved surface.

It has been shown above that this is the area of a rectangle whose adjacent sides are $2\pi r$ and h.

$\therefore$ **area of curved surface $= 2\pi rh$.**

$\therefore$ **total area of surface of cylinder**

is $$2\pi r^2 + 2\pi rh$$

or $$2\pi r(r + h).$$

172. Measurement of Volumes of Prisms.

It was stated above (§ 165) that the amount of space

FIG. 164.
Norman Porch, Canterbury.

[To face page 196.

occupied by a solid is called its volume. We must now examine how this volume is measured.

Unit of volume. The unit of volume is a **cubic inch,** *i.e.*, the volume of a cube each of whose edges is one **inch** in length, and each of its faces a square whose area is a **square inch.** For larger volumes we may use cubic feet or yards. In the metric system the commonest units are the cubic centimetre and the cubic metre.

173. Volume of a Prism.

In Fig. 166, *ABCD* represents a rectangle 4 in. by 3 in.

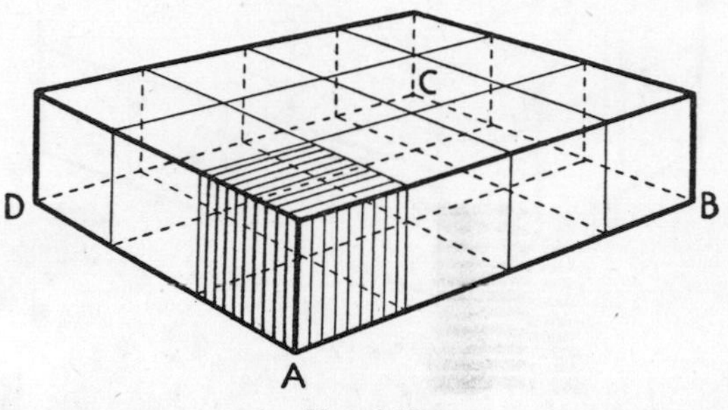

Fig. 166.

Its area is **12 sq. in.** On each of these square inches, a cubic inch is placed such as the one which is shaded. There are twelve of these cubic inches in all and the whole solid is a rectangular prism whose base is the rectangle *ABCD*, 4 in. by 3 in. and the height 1 in.

The volume of the prism is clearly 12 cub. in., or area of base (12 sq. in.) × height (1 in.).

If a similar prism of the same dimensions be placed on top of this the two together form a prism with rectangular base, **4 in. × 3 in.,** and height **2 in.**

The volume now = (4 in. × 3 in.) × **2 in.** = 24 cub. in.

If a third prism be added, as in Fig. 167,

the volume of the whole = (4 in. × 3 in.) × **3 in.**
= (area of base) × height.

A similar result will follow for any number of layers. In all cases it may therefore be concluded that

volume of prism = (area of base) × height.

In this explanation the lengths of the edges are an

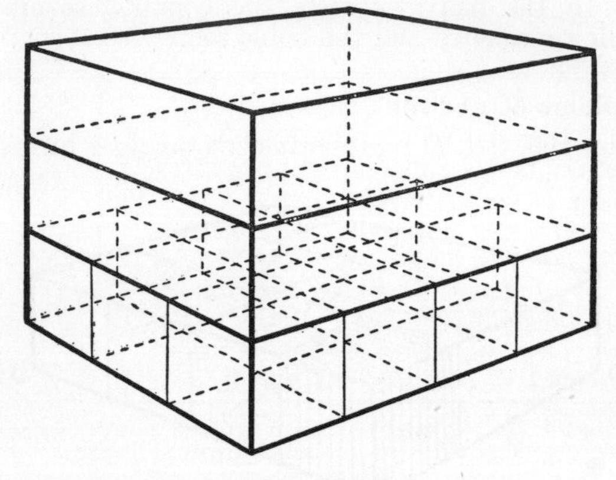

Fig. 167.

exact number of units, but, as was demonstrated in the case of the area of a rectangle (§ 91), the result may be shown to be true when the lengths involve fractions.

174. Prism Law of Volume.

The above rule can readily be shown to hold when the base is any other rectilineal figure. We may therefore deduce the rule for the determination of the volume of any right prism, viz.,

Volume of prism = area of base × height.

This is called the **prism law of volume.**

Since a cross-section perpendicular to the sides, *i.e.*, a normal section, is identically equal to the base, the prism law can be written thus:

Volume of prism = area of cross-section × height.

175. Volume of a Cylinder.

Let AB, Fig. 168, represent a side of a regular polygon inscribed in the circle PQ, which is the base of the cylinder depicted.

Let $ABCD$ be a lateral side of a right prism which is inscribed in the cylinder.

When the number of sides of the polygon which is the base of the prism is large, the straight line AB will be very nearly equal to an arc of the circle, and the volumes of the prism and cylinder will be nearly equal.

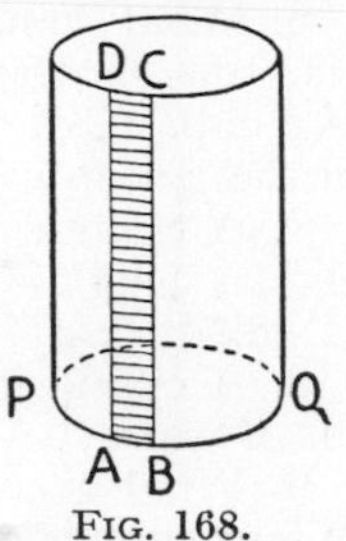

Fig. 168.

If the number of sides of the polygon inscribed in PQ be greatly increased, and consequently the number of lateral faces of the prism similarly increased, the volume of the prism will be approximately equal to that of the cylinder.

When the number of sides be increased without limit, we may conclude that—

Volume of cylinder = volume of prism.

∴ Using the volume law for Prisms (§ 174)

Volume of cylinder = area of base × height, *i.e.*, the prism law of volume holds for a cylinder.

Let r = radius of base of cylinder.
Then πr^2 = area of base or cross-section.
Let h = height of cylinder.

$$\therefore \textbf{volume of cylinder} = \pi r^2 h.$$

Exercise 20

1. Find the volume in cubic feet of a uniform rectangular beam, 18 ft. long, the area of whose normal cross-section is 40 sq. in.

2. A rectangular prism is made of metal and is 11 in. × 9 in. × 7 in. Find its weight in ounces if a cub. in. of the metal weighs 4·18 oz.

3. A uniform bar, rectangular in cross-section, is 3 ft. 8 in. long, and its cross-section is 2·5 sq. in. What is its volume?

4. A cub. ft. of lead is hammered out in order to make a square sheet, $\frac{1}{2}$ in. thick. What is the area of the square?

5. It is required to make 1000 cylindrical jam tins, 3·5 in. high and 3 in. in diameter. What is the total amount of tin required, if 10 per cent. is wasted in the cutting? (Take $\pi = 3·14$.)

6. What volume of jam will be contained by each of the tins in the previous question when each tin is full. If a cub. in. of the jam weighs 0·645 oz. what is the weight of the jam in each tin when it is full?

7. A cylindrical oil drum has a base of diameter 1·4 ft. and its height is 2 ft. How many gallons of oil will it hold? (1 gallon = 277·5 cub. in. approx.)

8. A cylindrical jar is 5 in. high, and it holds 30·5 cub. in. of water. What is the area of its cross-section?

9. What would be the cost of painting the curved surface of four cylindrical pillars, each 24 ft. high, and whose radius of cross-section is 9 in., at $2\frac{1}{2}d.$ per sq. ft.?

10. If the volume of a cylinder be $1\frac{1}{2}$ cub. ft. and its height be 3 ft., what is the radius of its cross-section?

11. In a hollow cylinder the circles of the cross-section are concentric. If the internal diameters of these circles be 2·2 in. and 3·8 in. respectively, and the height be 6·5 in., find the volume of the hollow interior.

CHAPTER 25

PYRAMIDS

176. Construction of a Pyramid.

In Fig. 169, $ABCD$ represents a square with its diagonals intersecting at O. OP is drawn perpendicular to the plane of $ABCD$. P is any point on this perpendicular and is joined to the points A, B, C, D.

The result is a solid figure bounded by a square as its

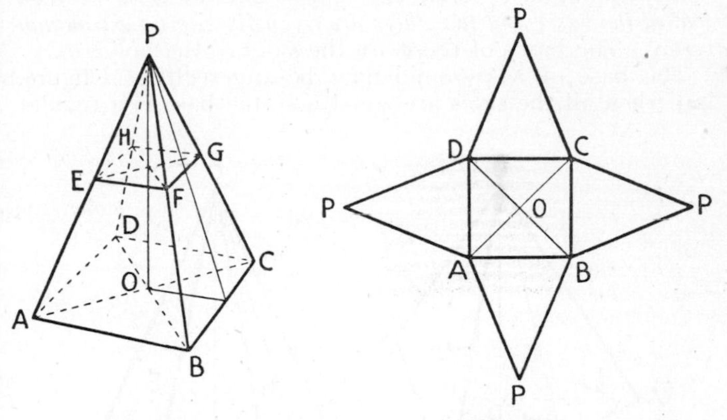

FIG. 169. FIG. 170.

base, and four triangles PAB, PBC, PCD, PDA with a common vertex P.

This solid is called a **pyramid**. OP is a **central axis** and the length of it is the height of the pyramid.

A section perpendicular to this axis, such as $EFGH$, is a similar figure to the base, *i.e.*, in this case, a square.

This is characteristic of all such pyramids. All normal cross-sections are similar figures.

A paper pyramid may readily be made by drawing a square on suitable paper, as $ABCD$, Fig. 170. On each side of the square construct isosceles triangles, all of the same height. The figure should then be cut out and the triangles folded about their bases and bent over till their vertices *P* come together. With a little ingenuity a method may be devised for keeping the slant edges together.

177. Regular Pyramids.

In the previous section we have seen how a pyramid may be constructed. We now proceed to give a formal definition of the solid.

Pyramid.

A pyramid is a solid, one of whose faces is a polygon (called the base) and the others are triangles having a common vertex. The bases of these are the sides of the polygon.

The base of a pyramid may be any rectilineal figure, but when all the sides are equal, *i.e.*, the base is a regular

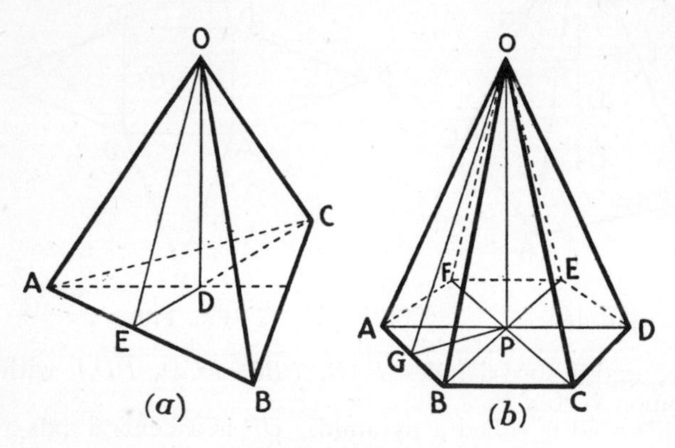

Fig. 171.

polygon, or an equilateral triangle, and the straight line joining the vertex to the centre of the base is perpendicular to the base, it is called a **regular pyramid**.

Such pyramids are named after the base. Thus, the pyramid of Fig. 169 is a square pyramid. If the base is an equilateral triangle it is called a **triangular pyramid,** or a **tetrahedron** (*i.e.*, having four faces), *see* Fig. 171 (*a*).

If the base is a hexagon (Fig. 171 (*b*)) it is a **hexagonal pyramid.** If the slant sides of a tetrahedron are also equilateral triangles it is called a **regular tetrahedron.** Fig. 171 (*a*) is an example of this particular form of a triangular pyramid.

178. The Cone.

A cone is a pyramid such that the base and every normal cross-section is a circle.

When the central axis, OP (Fig. 172 (*a*)), is perpendicular to the base the solid is called a **right cone.**

A cone can be constructed by cutting out a piece of paper in the shape of a sector, such as $PABC$ (Fig. 172 (*b*)) and

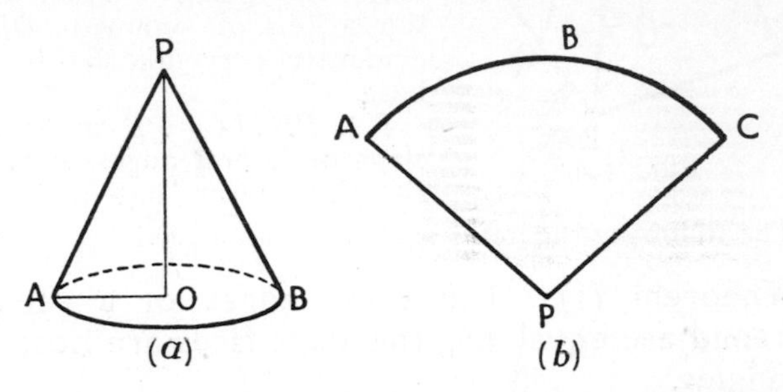

Fig. 172.

rolling together the radii PA and PC until C coincides with A, and PA with PC.

Thus the arc ABC becomes the circumference of the base of the cone and P becomes the vertex.

When the radius AP and the angle APC are known, the length of the arc ABC can be calculated (§ 123).

Hence, the length of the circumference of the base being known, the radius of the base, OA, is found.

AP is also known.

$$\therefore \text{ the height of the cone, } OP = \sqrt{AP^2 - OA^2}.$$

AP is called the **slant height** of the cone and the angle APB is the **vertical angle.**

179. Geometry of a Pyramid.

We next proceed to consider some of the geometric relations between edges, sides, etc., of a pyramid, basing the conclusions upon the definition of a pyramid as given in § 177.

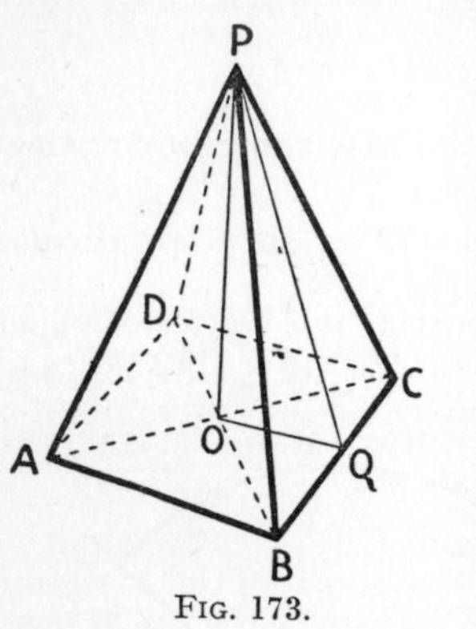

FIG. 173.

Fig. 173 represents a square pyramid, $ABCD$ being the square base and O, the intersection of its diagonals, thus being the centre of the base. OP is the axis of the pyramid and represents the height. From considerations of symmetry OP is evidently perpendicular to the base.

PA, PB, PC, PD are called **slant or lateral edges,** and AB, BC, CD, DA are base edges.

The following theorems may now be proved:

Theorem (I). The slant edges of a regular pyramid are equal, *i.e.*, the slant faces are isosceles triangles.

Consider the slant face PBC (Fig. 173).

Now, OP is perpendicular to the plane of the base.

$\therefore$ it is perpendicular to every straight line it meets in that plane (*see* § 29).

$\therefore$ $\angle$s POB, POC are right angles.

Since the diagonals of a square bisect each other

$$\therefore \quad OB = OC$$

$$\therefore \quad PO^2 + OB^2 = PO^2 + OC^2,$$

i.e.,

$$PB^2 = PC^2$$

$$\therefore \quad PB = PC$$

and the $\triangle PBC$ is isosceles.

Let Q be the mid-point of BC. Join OQ.
Then PQ is called the **slant height of the pyramid.**

Theorem (2). The slant height of a pyramid is perpendicular to a base edge.

In $\triangle$s PQB, PQC:

(1) $PB = PC$ (proved above)
(2) $BQ = QC$ (construction)
(3) PQ is common.

$\therefore$ **$\triangle$s PQB, PQC are congruent.**
In particular $\angle PQB = \angle PQC$.
$\therefore$ **PQ is perpendicular to BC.**

(Compare § 62.)

Definition :

The angle PQO is **the angle between the lateral face PBC and the base ABCD** (§ 30).

By the use of the above theorems and the Theorem of Pythagoras the slant height and slant edge of a square pyramid may be found when the base and height are known.
Similar relations also hold for other regular pyramids.

180. Area of the Surface of a Right Pyramid.

The total surface of any pyramid consists of (1) the area of the base together with (2) the sum of the areas of the lateral faces.

(1) **Area of the base.** The base is a regular rectilineal figure the area of which can be found by using rules previously considered.

(2) **Area of lateral faces.** As shown in § 179 the lateral faces are equal isosceles triangles.

In each of these it has been proved that the slant height, as PQ in Fig. 173, is the height of the corresponding triangle.
Considering the $\triangle PBC$,

Area of $\triangle = \frac{1}{2}BC \times PQ$
$= \frac{1}{2}$ **base edge $\times$ slant height.**

In the same way the areas of other faces may be found.

$\therefore$ sum of areas of lateral faces

$= \frac{1}{2}$(sum of sides of base $\times$ slant height)

$= \frac{1}{2}$(perimeter of base) $\times$ (slant height).

181. Area of the Surface of a Cone.

As with pyramids, the total area of surface of a cone is the sum of (1) area of circular base and (2) area of the curved surface.

To find the area of the curved surface.

Let QR (Fig. 174) be a side of a regular polygon with a large number of sides, inscribed in the circle which is the base of the cone.

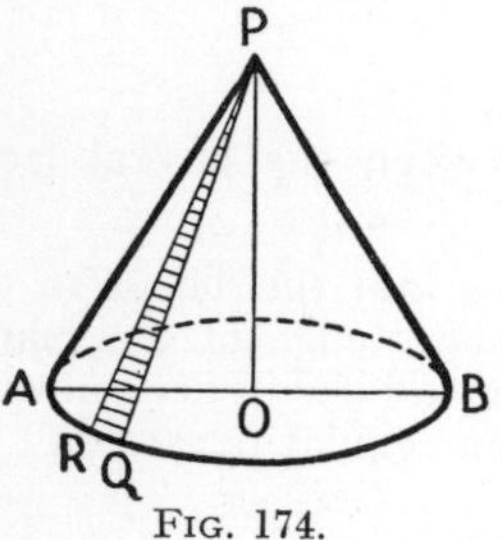

Fig. 174.

Q and R being joined to P, PQR may be regarded as one of the lateral faces of a pyramid of which QR is one of the sides of the base.

Suppose the number of sides of the polygon to become very large, so that QR becomes very small. It will then not differ appreciably from the corresponding arc of the surrounding circle.

Let a perpendicular be drawn from P to the middle of PQ.

The length of this will be very nearly equal to the slant height of the cone.

Then area of $\triangle PQR$ is approximately equal to

$$\frac{1}{2}QR \times \text{(slant height of cone)}.$$

$\therefore$, as in § 180,

Total area of lateral faces of pyramid equals

$$\frac{1}{2}\text{(perimeter of base)} \times \text{(slant height)}.$$

If the number of sides of the polygon be increased without limit:

(1) the perimeter of the base of pyramid is equal to the circumference of the base of the cone;

(2) the perpendicular from P to QR is indistinguishable from the slant height of the cone.

(3) the lateral surface of the pyramid is equal to the curved surface of the cone.

$\therefore$ **Area of curved surface of cone equals**

$$\tfrac{1}{2}\text{ (circum. of base)} \times \text{(slant height)}.$$

Let r = radius of base.
l = slant height.
h = height.

Then $l = \sqrt{r^2 + h^2}$.

$$\text{Area of curved surface of cone} = \tfrac{1}{2}(2\pi r \times l) = \pi r l = \pi r\sqrt{r^2 + h^2}.$$

$$\therefore \textbf{ total surface of cone} = \pi r^2 + \pi r l = \pi r(r + l) = \pi r(r + \sqrt{r^2 + h^2}).$$

182. Volume of a Pyramid.

Fig. 175 represents a cube and its diagonals AF, BG, CH, DE are drawn intersecting at O.

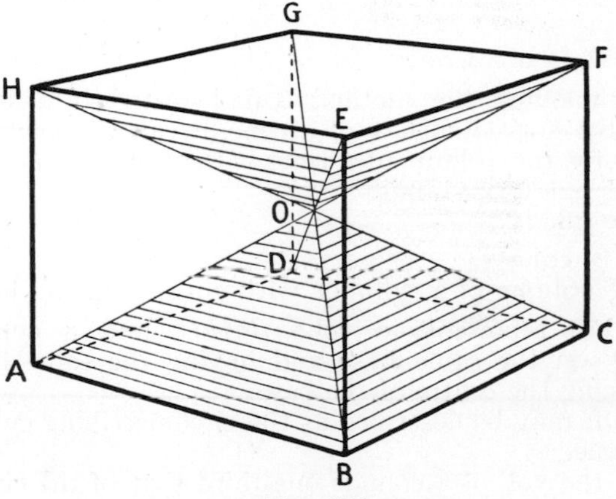

FIG. 175.

Symmetry suggests that they are concurrent. These diagonals form the slant edges of **six** pyramids, each of which has its vertex at O, and one of the faces of the cube as a base, as for example the pyramid $OABEH$. The base of this pyramid is the face $ABEH$, and the slant edges are OA, OB, OE, OH.

The six pyramids are clearly all equal in volume. Their bases being equal and the height of each being the perpendicular (not shown) drawn from O to the centre of the base; the length of this is half that of the side of a square.

$\therefore$ **the volume of each pyramid is one-sixth of the volume of the whole cube.** Thus, **it is one-third of the volume of half the cube,** *i.e.*, of the rectangular prism whose base is a face of the cube, such as $ABEH$, and whose height is the same as that of the pyramid.

Thus, **the volume of each pyramid is one-third of a prism of the same base and the same height,**

***i.e.* volume of pyramid $= \frac{1}{3}$(area of base) $\times$ height.**

We have been dealing with a particular case of a square pyramid, but the rule can be shown to hold for any pyramid.

183. Volume of a Cone.

By employing the method demonstrated in § 181 for finding the curved surface of a cone, it can be shown that the rule for the volume of a cone is the same as the corresponding one for pyramids.

$\therefore$ the rule is

volume of cone
$= \frac{1}{3}$ volume of a cylinder with equal base and height.

Fig. 176 represents a cone, $PABC$, and a cylinder, $EABCD$, on the same base and having the same height OP, O being the centre of the base.

$ABCDE$ may be described as the circumscribing cylinder of the cone.

Thus, the vol. of a cone is one-third that of the circumscribing cylinder.

Let r = radius of base
h = height of cone.

Volume of circumscribing cylinder $= \pi r^2 h$.

$\therefore$ Volume of cone $= \frac{1}{3}\pi r^2 h$.

184. Frustum of a Pyramid or Cone.

Fig. 177 represents a cone in which *DE* is a section parallel to the base.

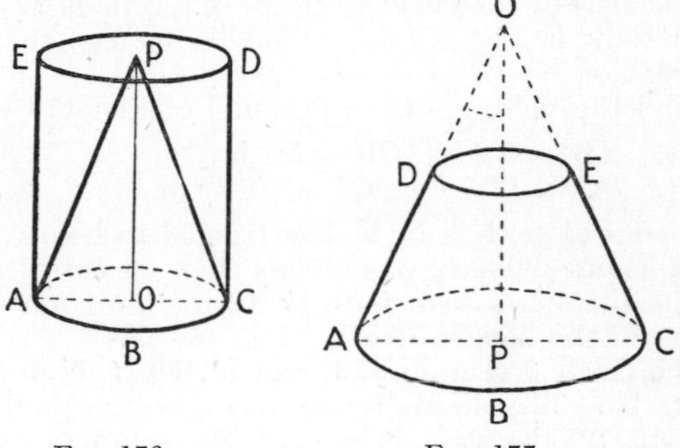

FIG. 176. FIG. 177.

If the part of the cone above this be removed the remaining solid *DABCE* is called a **frustum of the cone.**

Similarly, in Fig. 169, if that part of the pyramid above the section *EFGH* be removed the remainder is a **frustum of the pyramid.** So with other pyramids.

The volume of a frustum can be obtained as the difference between the volume of the complete solid and the part removed.

The top part of a funnel is an example of a frustum in everyday life. Among many other examples are a bucket, a flower-pot and many lamp-shades.

Exercise 21

1. A pyramid 12 in. high stands on a square base of 6-in. side. Find (*a*) its volume, (*b*) its total surface area.

2. Find the volume of a hexagonal pyramid, each side of the hexagon being 1·5 in., and the height of the pyramid 8 in.

3. Find (*a*) the total surface area, and (*b*) the volume, of a cone of height 9 in. and radius of base 4·5 in.

4. The area of the curved surface of a cone is 22·4 sq. in. and the slant height is 8 in. Find the area of the base of the cone.

5. Find the volume of the pyramid of Fig. 173 when

   (1) $AB = 3{\cdot}2$ in., $OP = 5{\cdot}1$ in.
   (2) $OQ = 11{\cdot}7$ cm., $OP = 10{\cdot}8$ cm.

6. A conical tent is to be constructed to house 10 men, each of whom must have not less than 60 cub. ft. of air. If the height of the tent is to be 12 ft., what must be the diameter of the base?

7. The Great Pyramid of Egypt is 450 ft. high and has a square base of side approximately 746 ft. Find (1) the slant edge, (2) the slant height, (3) the volume.

8. Each of the sides of the base of a regular hexagonal pyramid is 2 in. long, and the height of the pyramid is 4·5 in. Find, (1) the slant height, (2) the total surface area of the pyramid.

9. A cylindrical column 4 ft. in diameter and 6 ft. high is surmounted by a cone of the same width and 3 ft. high. Find the area of sheet metal required to cover the whole lateral surface.

10. In a cone whose vertical angle is 60°, two parallel sections are drawn perpendicular to the axis and at distances of 3 in. and 5 in. from the vertex. Find (*a*) the area of the curved surface, (*b*) the volume of the frustum so formed.

11. In a square pyramid of base 4-in. side and height 6 in., a section parallel to the base is made half way between the base and the vertex. Find the area of the surface of the frustum thus cut off. Find also its volume.

CHAPTER 26

SOLIDS OF REVOLUTION

185. The Cylinder.

If the cover of this book, or a door, be rotated about one edge which is fixed, every point on the edge which rotates will trace, in space, a circle or an arc of a circle, the centre of which is on the fixed edge. Since all points on the moving edge are the same distance from the fixed edge, all the circles will have equal radii. Considering the rotation of the whole plane surface of the cover, since it is rectangular in shape, a portion of a cylinder will be marked out in space, or a complete cylinder, if there is a complete rotation.

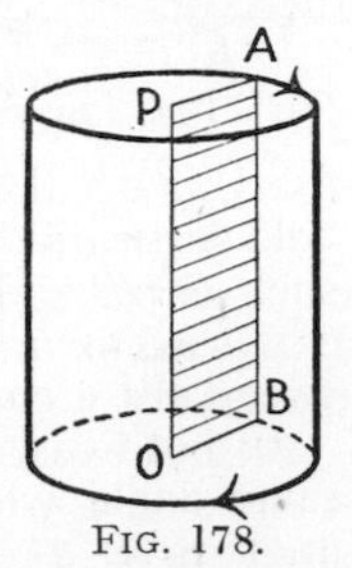

FIG. 178.

Treating this more generally, in Fig. 178, *OPAB* represents a rectangle with one side *OP* fixed. This rectangle rotates round *OP* as an axis of rotation. Every point on *AB* describes a circle and all such circles, having equal radii, are equal in area. Consequently a cylinder is marked out in space.

A solid which is thus described by the rotation of a line or figure about a fixed axis is called a solid of revolution.

The fixed straight line, *OP* in Fig. 178, about which the rotation takes place is called the **axis of rotation.**

A straight line which thus rotates (as *AB* in Fig. 178) is called a **generating line.**

186. The Cone as a Solid of Revolution.

Take a set square, preferably the 90°, 60°, 30° one, and stand it vertically upright with the shortest side on a piece of paper. Holding it firmly upright by the vertex, rotate it round the other side containing the right angle, *i.e.*, the side opposite to 60°. If this be represented by $\triangle OPA$ in

Fig. 179 it will be seen that as this rotates about *OP* as an axis of rotation, *A*, in a complete revolution, will describe the circumference of a circle *ABC*. Every point on *PA* will also describe a circle with its centre on *OP*. Consequently **in a complete rotation a cone will be marked out, in space**; *OP*, the axis of rotation, will be the central axis of the cone, and *ABC* will be its base. The line *AP*, *i.e.*, the hypotenuse, is the **generating line.**

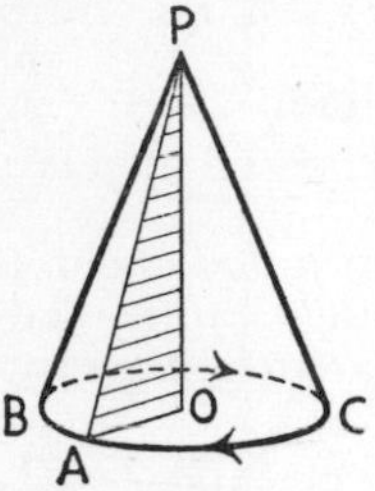

FIG. 179.

As an additional illustration, let the pencil of a compass be extended, so that the arm which holds it is longer than the other. If the pointed end be fixed, as usual, into a horizontal piece of paper, and the arm held vertically, the two arms will form two sides of a triangle, such as *OP* and *PA* in $\triangle OPA$ (Fig. 179). On rotating the compass, as usual keeping *OP* vertical, the pencil arm will sweep out a cone, as in the previous experiment.

In both cases the **generating line, AP, always makes the same angle with OP.** In the case of the set square this angle is 30°.

As a practical example, the arm of a crane, when transferring a suspended load from one point to another, rotating about a fixed position on the base, marks out a cone in space. This time the conc is upside down compared with the previous examples. It will be as *AOA'* in Fig. 180.

187. The Double Cone.

A general treatment of this aspect of the formation of a cone as a solid of revolution is as follows.

In Fig. 180 let *PQ* be a fixed straight line of indefinite length.

Let *AB* be another straight line, also of indefinite length, intersecting *PQ* at *O* and making with it the angle *AOQ*. Now, suppose *AB* to rotate in space around *PQ*, so that it continues to intersect it at *O* and the **angle AOQ remains constant for any position of AB.** Any points on *AB* such as C and D will trace out in space circumferences of circles as AB rotates about PQ.

After half a complete rotation AB will be once more in the plane from which it started in the position $A'B'$. C and D will be at C' and D' respectively and every point on AB will have marked out a semicircle.

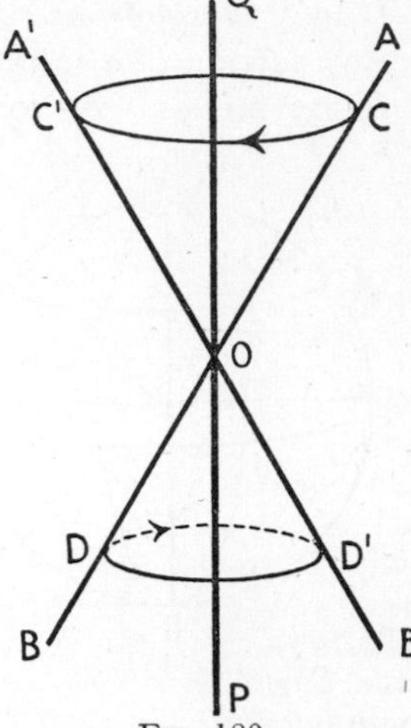

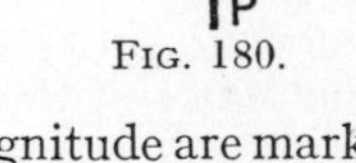
FIG. 180.

Continuing the rotation, AB, after another half rotation, will be back in its original position in the plane. C and D will be back to their original positions and will have described circles with CC' and DD' as diameters, the planes of which will be perpendicular to PQ.

All points on AB will thus describe circles and the complete solid so formed will be a **double cone** with a **common apex at O.**

The cones as described are of indefinite size, since the straight lines AB and PQ are of unlimited length, but cones of definite magnitude are marked out when a distance on AB is fixed, such as COD. The glass often used when eggs are boiled is an example of a double cone.

188. The Sphere.

The two diameters AB, PQ of the circle $APBQ$ (Fig. 181) are perpendicular to one another, O being the centre.

Let this circle rotate completely in space round AB as an axis of rotation.

In a half rotation P will rotate to the position of Q, and any point C to C'. Continuing the rotation P and C will return to their original positions and the completely enclosed solid, known as a ball or **sphere**, will have been generated. Any point on the circumference $ACPB$ will describe a circle whose plane is perpendicular to AB.

$\therefore$ any section perpendicular to AB is a circle.

Similarly the circle could be rotated about **any** other diameter. The same sphere would be formed and in each of these every section perpendicular to the diameter is a circle.

It may therefore be concluded that,

Any section of a sphere is a circle.

Great circles. A section which passes through the centre of a sphere is called a great circle.

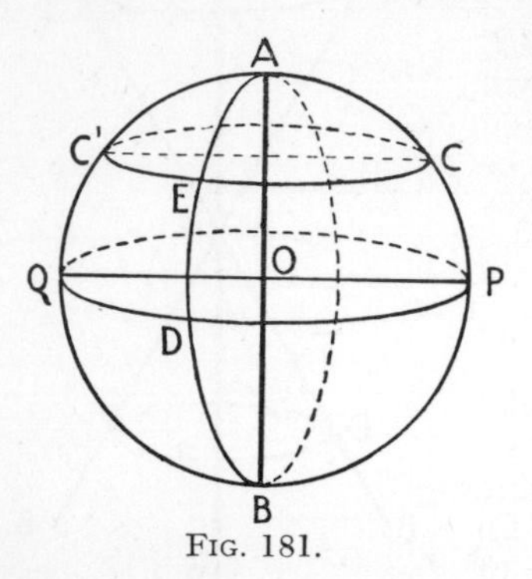

FIG. 181.

Thus, in Fig. 181 *PDQ* and *ADB* are great circles.

The radius of a great circle is always equal to that of the sphere.

Small circles. All other sections of a sphere which are not great circles are called small circles. In Fig. 181, *CEC'* is an example. The radius of a small circle is always less than that of the sphere and varies according to the distance of the section from the centre of the sphere.

189. The Earth as a Sphere.

The earth is approximately a sphere, rotating completely on an axis every twenty-four hours. It is not an exact sphere, being slightly flattened at the ends of its axis of rotation. These ends are termed the **North and South Poles** (*see* § 32).

Fig. 182 represents the earth as a sphere, centre *O*, *NOS* being the axis of rotation.

∴ *N* represents the North Pole and *S* the South Pole.

The circle *EABW* represents a great circle, perpendicular to *NS* and halfway between *N* and *S*. It is known as the **equator.**

ON and *OS* represent the north and south directions from *O* (*see* § 32) and *OE* and *OW* represent the east and west directions.

CGHF is a small circle perpendicular to the axis and therefore parallel to the plane of the equator.

NGBS and *NHAS* are great circles passing through the

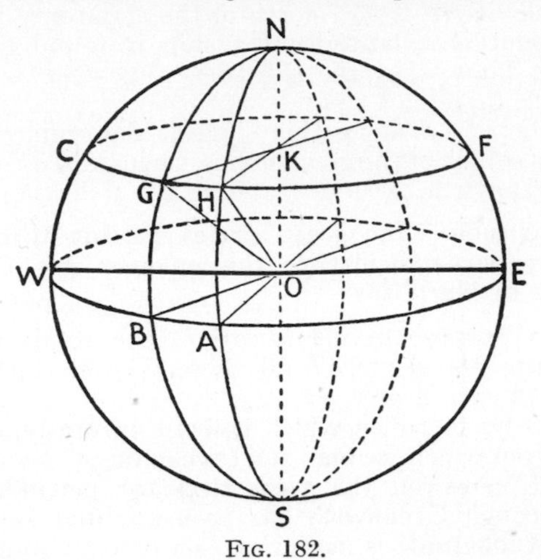

FIG. 182.

poles and therefore perpendicular to the planes of the equator and of the small circle *CGHF*.

190. Determination of Position on the Earth's Surface.

The position of a point such as *G*, on the earth's surface, is clearly determined by the intersection of the great circle, *NGBS*, and the small circle, *CGHF*. If these are known the position of *G* is known. The problem is, how are these circles to be identified, on a map or chart?

(1) **Latitude.** The quadrant arc *NGB* subtends a right angle, *NOB*, at the centre of the sphere. As stated in § 22, this can be divided into 90 degrees. Corresponding to each degree on it small circles may be drawn in planes parallel to the equator.

These small circles are called circles of latitude.

Since an arc of a circle is proportional to the angle which it subtends at the centre of the circle (§ 120), if the angle

BOG can be determined the number of degrees in *BG* is known, and the circle of latitude is known. Thus if the angle *BOG* be 55°, then the latitude of every point on the small circle *CGHF* is 55°, north of the equator.

If therefore the latitude of a ship is found to be 55° north, we know that the ship lies somewhere along the small circle *CGHF*.

Just where it is along this circle is determined if it be known on which of the great circles, such as *NGBS*, it is also placed. This will be investigated in the following section.

(2) **Longitude. The great circles passing through the poles and perpendicular to the equator are circles of Longitude or Meridians.**

For the purpose of identification, the semi-equator is divided into 180 degrees, and there is a Meridian corresponding to each degree.

There is **no meridian which is fixed** naturally, as is the equator from which we may start reckoning. By universal agreement, however, the **great circle or meridian which passes through Greenwich** has been adopted as the zero circle, and longitude is measured east or west from it. If, for example, in Fig. 182, the great circle *NHAS* is the Meridian of Greenwich and the angle subtended by the arc *AB* is 15°, then the longitude of *G* is 15° west, since the angle *GKH* which measures the arc *GH* is equal to the angle *BOA*, each of these being the angle between the planes of the two circles (§ 27).

Consequently, **the position of G is 55° north latitude, and 15° longitude west of Greenwich.** Thus the position of *G* on a chart can be fixed.

As the distances involved are very great, for accurate measurements, each degree is divided into 60 minutes and each minute into 60 seconds as in § 22.

The actual determination of the latitude and longitude lies beyond the scope of this book.

191. Distances Measured on a Sphere.

Suppose it is required to determine the distance between the two points *B* and *G* in Fig. 182. A ship sailing on the sea between them could do so by a variety of paths. The **shortest distance between them, however, is the length of the arc between them on the great circle which passes through them.** In Fig. 182 the arc GB is the shortest distance between B and G.

The great circle, in this matter, corresponds to the straight line joining points on a plane (*see* § 6, Fig. 4).

This is an important matter for the navigator, whether on the sea or in the air. Before the distance BG can be determined we must clearly know the length of the great circle $NGBS$. This is the circumference of a circle whose diameter is that of the earth itself. We may use the rule of §§ 122 and 123, and thus find the arc BG.

192. Surface and Volume of a Sphere.

The methods by which formulæ are found for the surface and volume of a sphere require a greater knowledge of mathematics than is assumed in this volume.

The formulæ are therefore offered without proof.

(1) **Area of surface of a sphere.**

Let r = radius of sphere.

Then **area of surface** $= 4\pi r^2$.

(2) **Volume of a sphere.**

$$\text{Volume of sphere} = \tfrac{4}{3}\pi r^3.$$

193. Areas and Volumes of Cylinder, Cone and Sphere.

The following connections exist between the areas and volumes of the above solids in which the diameter of the sphere is equal to the diameters of the base of the cylinder and cone and to their heights.

The three solids are shown in Fig. 183, the sphere being inscribed in the cylinder, *i.e.*, the bases of the cylinder and cone, and the curved surface of the cylinder touch the sphere. They are therefore **tangential to the sphere.**

Let r = radius of sphere.

Then $2r$ equals (1) diameter of base of cylinder and cone, and (2) height of cylinder and cone.

Areas of curved surfaces of sphere and cylinder.

(1) Of cylinder $= 2\pi r \times 2r = 4\pi r^2$.
(2) Of sphere $= 4\pi r^2$.

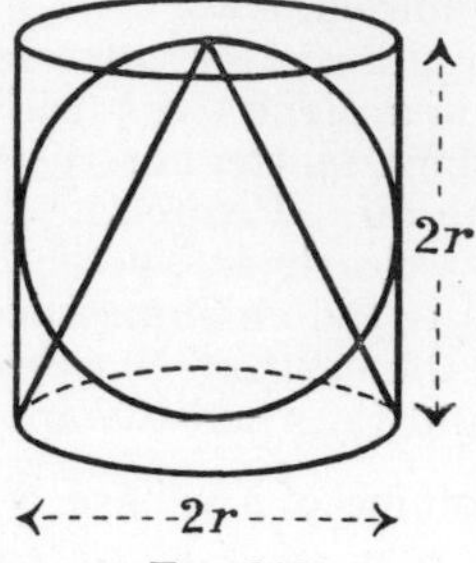

FIG. 183.

I.e., **Areas of curved surfaces of cylinder and inscribed sphere are equal.**

Volumes.

(1) Of cylinder $= \pi r^2 \times 2r = 2\pi r^3$.
(2) Of sphere $= \frac{4}{3}\pi r^3$.
(3) Of cone $= \frac{1}{3}\pi r^2 \times 2r = \frac{2}{3}\pi r^3$.

$\therefore$ ratio of volumes $= 2 : \frac{4}{3} : \frac{2}{3} = 6 : 4 : 2$
$= 3 : 2 : 1$.

Thus, the **volume of the cylinder** $=$ **sum of volumes of inscribed sphere and cone.**

Exercise 22

1. What is the area of the cloth required to cover a tennis ball of diameter $2\frac{1}{2}$ in.? Find also the volume of the ball.

2. Find the cost of gilding the surface of a spherical ball of radius 4 ft. at 1*s*. $4\frac{1}{2}d$. per sq. ft. $(\pi = \frac{22}{7})$.

3. What is the area of the earth's surface, assuming it to be a sphere of radius 3960 miles $(\pi = \frac{22}{7})$? Give the answer correct to the nearest 10 million sq. miles.

4. Find the ratio of the surface of a sphere to the surface of the circumscribing cube.

5. Find the area of the whole surface of a hemisphere of diameter 10 in. (Take $\pi = 3{\cdot}1416$.)

6. A cast-iron dumb-bell consists of two spheres of $2\frac{1}{2}$ in. diam., connected by an iron cylinder 6 in. long and 1 in. diameter. Find its weight if 1 cub. in. weighs 0·28 lb.

7. If a right-angled triangle be rotated about its hypotenuse, what is the solid formed?

8. An equilateral triangle of side 3 in. is rotated about one of its sides. Find the volume of the solid which is formed in a complete rotation.

9. A sphere of radius 5 in. exactly fits into a cubical box. Find the volume of the space which is unoccupied in the box.

PART II

FORMAL GEOMETRY

INTRODUCTION

As was stated on p. xvii, Part II of this book is designed to provide a short course in formal or abstract geometry. The theorems which comprise it are arranged so that their sequence provides a logical chain in which all geometrical facts which are employed in a proof have previously been proved to be true. In Part I, when theorems were proved, appeal was frequently made to intuitive reasoning or to conclusions which emerge from the considerable body of geometric knowledge which is the common heritage of modern civilisation.

In a course of formal geometry it is of the first importance that we should be scrupulously accurate as to the meanings of any terms which may be employed. Hence the importance of clear, precise definitions, as stated in Part I (§ 4). These definitions, together with a small number of axioms, constitute the starting point for logical mathematical reasoning, and should precede the formal study of the subject. But as these have already been stated and discussed in Part I they will not be repeated now, though it is desirable in a few cases to remind the student of them by quoting some of them.

Some of the proofs which will appear in Part II, have been given, substantially, in Part I, but are repeated so that they may take their logical position in the chain of theorems which constitute the system of geometric reasoning.

Constructions which appeared in Part I will not be found in Part II. A number of new ones, however, which are dependent for their proof upon theorems in Part II, are included.

The student after studying a theorem is strongly urged to test his mastery of it by reproducing the proof from memory. It is also advised that he should attempt to solve the exercises, or " riders ", which will be found at the end of each section.

A few of the proofs of theorems, such as Nos. 1–3, hardly seem to merit inclusion, as their truth is apparent, especially since they have been studied in Part I, but they are included so that the sequence of proofs may be complete.

SECTION I

ANGLES AT A POINT

Theorem I

If one straight line meets another straight line the sum of the two adjacent angles on one side of it is two right angles.

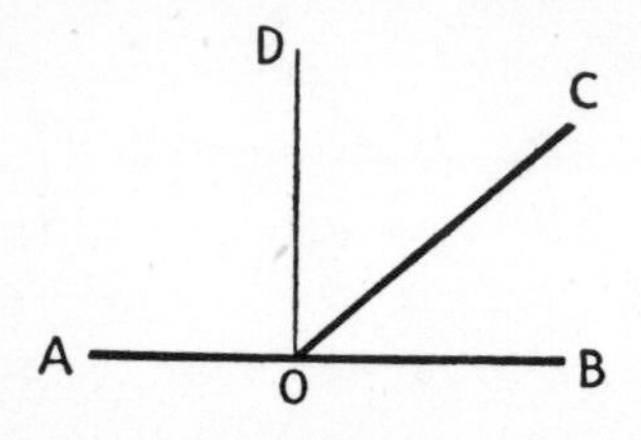

FIG. 184.

Given. OC meets the straight line AB at O.

To prove. $\angle AOC + \angle BOC =$ two right $\angle$s.

Construction. From O draw OD perpendicular to AB.

Proof. $\angle AOC = \angle AOD + \angle COD$.

Adding $\angle BOC$ to each,

$$\angle AOC + \angle BOC = \angle AOD + \angle COD + \angle BOC.$$

But $\angle COD + \angle BOC = \angle BOD$

$$= \text{a right } \angle.$$

$$\therefore \quad \angle AOC + \angle BOC = \angle AOD + \angle BOD,$$

i.e., $\angle \text{AOC} + \angle \text{BOC} =$ two right $\angle$s.

Theorem 2

(*Converse of Theorem* 1)

If at a point in a straight line two other straight lines on opposite sides of it make the two adjacent angles together equal to two right angles, these two straight lines are in the same straight line.

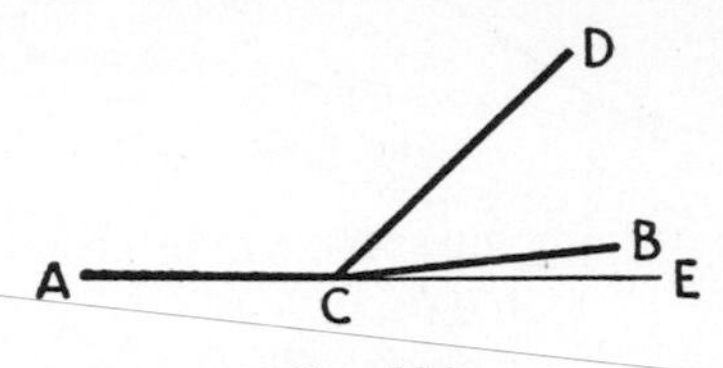

FIG. 185.

Given. CD meets the straight lines AC and CB at C, so that

$$\angle ACD + \angle BCD = \text{two right } \angle\text{s}.$$

To prove. AC and CB are in the same straight line.

Construction. Produce AC to any point E.

Proof. Since CD meets the straight line AE at C,

$\angle ACD + \angle DCE =$ two right $\angle$s (*Th.* 1)

but $\angle ACD + \angle DCB =$ two right $\angle$s (*given*).

$\therefore\ \angle ACD + \angle DCE = \angle ACD + \angle DCB.$

Subtracting the $\angle ACD$ from each side.

$$\therefore\ \angle DCE = \angle DCB.$$

$\therefore$ CE and CB must coincide,

but ACE is a straight line (*constr.*)

$\therefore$ **ACB is a straight line.**

Theorem 3

If two straight lines intersect, the vertically opposite angles are equal.

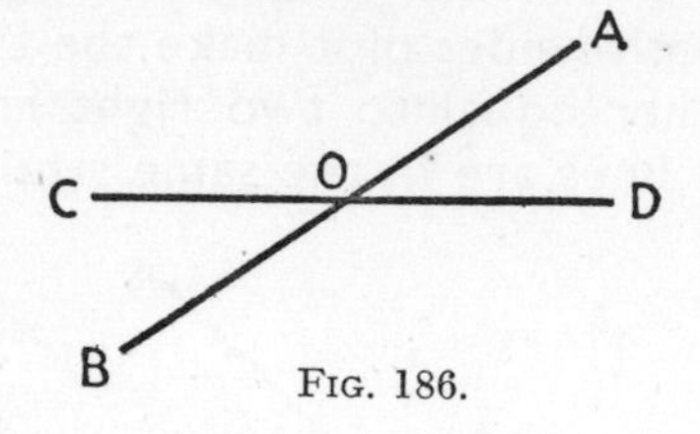

Fig. 186.

Given. The straight lines AB, CD intersect at O.

To prove. $\angle AOD = \angle BOC$

and $\angle AOC = \angle BOD$.

Proof. Since AO meets CD at O,

$$\therefore \quad \angle AOC + \angle AOD \angle 2 \text{ right } \angle\text{s} \quad (\textit{Th. } 1)$$

and since CO meets AB at O,

$$\therefore \quad \angle AOC + \angle BOC = 2 \text{ right } \angle\text{s} \quad (\textit{Th. } 1)$$

$$\therefore \quad \angle AOC + \angle AOD = \angle AOC + \angle BOC.$$

From these equals take away $\angle AOC$.

$$\therefore \quad \angle \mathrm{AOD} = \angle \mathrm{BOC}.$$

Similarly it may be proved that

$$\angle \mathrm{AOC} = \angle \mathrm{BOD}.$$

Exercise 23

1. In Fig. 186 prove that the bisectors of the angles BOD, AOD are at right angles.
2. In Fig. 186 prove that the straight lines which bisect the angles AOC, BOD are in the same straight line.
3. An angle AOB is bisected by OC. CO is produced to D and AO is produced to E. Prove that the $\angle COB = \angle DOE$.
4. The line OX bisects the angle AOB, XO is produced to Y. Prove $\angle AOY = \angle BOY$.

SECTION 2

CONGRUENT TRIANGLES. EXTERIOR ANGLES

Theorem 4

Two triangles are congruent if two sides and the included angle of one triangle are respectively equal to two sides and the included angle of the other.

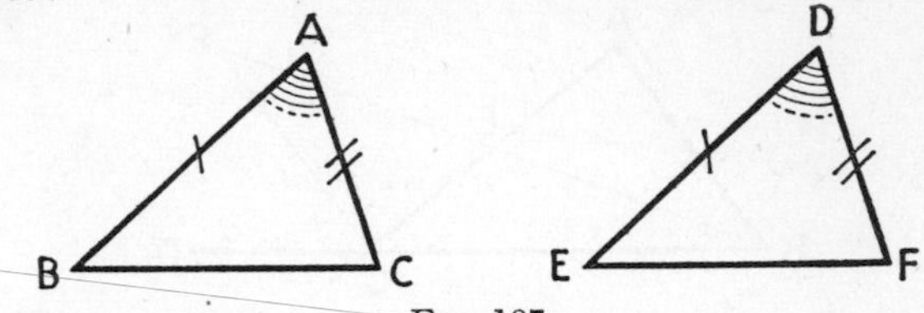

FIG. 187.

Given. ABC, DEF are two triangles such that

$$AB = DE$$
$$AC = DF$$
$$\text{included } \angle BAC = \angle EDF.$$

To prove. Δs are congruent.

Proof. Apply the ΔABC to the ΔDEF so that

(1) The point A falls on D.
(2) AB lies along DE.

Since $AB = DE$ (*given*)

$\therefore$ the point B falls on E.

Since AB lies along DE, and $\angle A = \angle D$,

$\therefore$ AC must lie along DF.

And since $AC = DF$,

$\therefore$ the point C falls on F.

Since only one straight line can join two points,

$\therefore$ BC coincides with EF.

$\therefore$ all the sides of ΔABC coincide with the corresponding sides of ΔDEF.

$\therefore$ **ΔABC is congruent with ΔDEF.**

Note.—This method of proof is called " superposition "—*i.e.*, we test if two figures are congruent by applying one to the other.

Modern mathematicians have raised objections to this as a method of proof. However, no other satisfactory method of proving this theorem has been evolved.

Theorem 5

If one side of a triangle be produced, the exterior angle so formed is greater than either of the interior opposite angles.

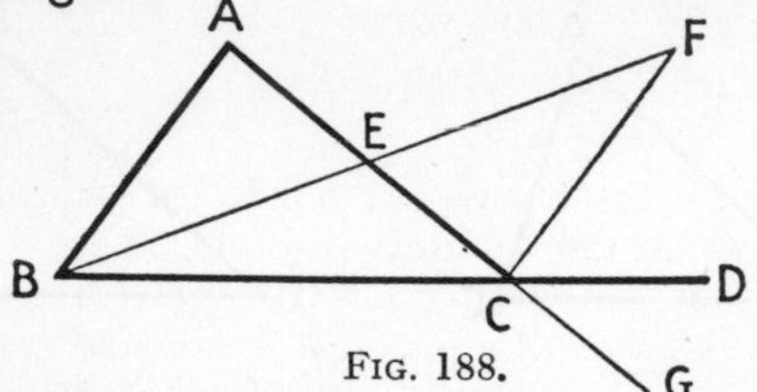

FIG. 188.

Given. In $\triangle ABC$, BC is produced to D.

To prove. Ext. $\angle ACD$ is greater than either of the interior opposite $\angle$s ABC or BAC.

Construction. Let E be the mid point of AC. Join BE.
Produce BE to F making $EF = BE$.
Join FC.

Proof. In $\triangle$s ABE, FCE :

(1) $AE = EC$ (*constr.*)
(2) $BE = EF$ (*constr.*)
(3) $\angle AEB = \angle FEC$ (*Th.* 3)

$\therefore$ $\triangle$s are congruent (*Th.* 4)

In particular $\angle EAB = \angle ECF$.
But $\angle ACD > \angle ECF$.
$\therefore$ $\angle ACD > \angle BAC$.
To prove $\angle ACD > \angle ABC$.

If AC be produced to G and BC bisected, it can similarly be proved that

ext. $\angle BCG >$ int. $\angle ABC$.
But $\angle BCG = \angle ACD$ (*Th.* 3)
$\therefore$ $\angle ACD > \angle ABC$.

$\therefore$ $\angle ACD$ is greater than either of the interior opposite angles

Exercise 24

1. ABC is an isosceles $\triangle$ of which A is the vertex. BA and CA are produced to P and Q respectively so that $AQ = AP$. Join BQ and CP. Prove that these straight lines are equal.

2. OA and OB are two equal straight lines and OC bisects the angle between them, X is any point on OC. Prove that $XA = XB$.

3. In Fig. 188 (*Th.* 5) prove

   (1) $\angle BEC > \angle BAC$.
   (2) $\angle FCD > \angle EFC$.

4. Prove that from a point outside a straight line only one perpendicular can be drawn to the line.

5. Prove that the diagonals of a square are equal.

6. The mid-points of the sides of a square are joined up in succession to form a quadrilateral. Prove that this quadrilateral is a square.

SECTION 3

PARALLELS

Definition. *Parallel straight lines are such that, lying in the same plane, they do not meet however far they may be produced in either direction.*

Note.—The method of proof employed in Theorem 6 which follows is known as "*Reductio ad absurdum*". This form of proof assumes that the theorem to be proved is untrue. When this leads to a conclusion which is either geometrically absurd or contrary to the data, it follows that the assumption cannot be true. Consequently the truth of the theorem is established.

Theorem 6

If a straight line cuts two other straight lines so that the alternate angles are equal, then the two straight lines are parallel.

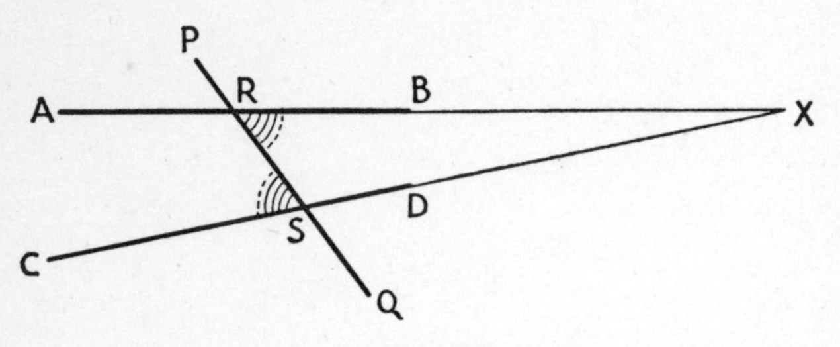

FIG. 189.

Given. The straight lines AB, CD are cut by a transversal PQ at R and S and

$$\angle BRS = \angle RSC \text{ (alternate angles, § 54).}$$

To prove. AB and CD must be parallel.

Proof. If AB and CD are not parallel, then, if produced in either direction, they must meet. (*Def.*).

Let them be produced towards B and D so that they meet in X.

Then XRS is a triangle.

One of its sides XS is produced to C, and $\angle RSC$ is an exterior angle of the $\triangle$.

$$\therefore \quad \angle RSC > \text{interior } \angle XRS \quad (Th.\ 5)$$

But we are given that $\angle XRS = \angle RSC$.

$\therefore$ the assumption that the straight lines AB and CD will meet has led to a conclusion which is contrary to the hypothesis, viz., $\angle BRS = \angle RSC$.

$\therefore$ it cannot be true that, with this hypothesis, AB and CD will meet when produced towards B and D.

In the same way it can be shown that the straight lines will not meet when produced towards A and C.

$\therefore$ since they will not meet when produced in either direction, by definition

AB and CD are parallel.

Theorem 7 ✓

If a straight line cuts two other straight lines so that

(1) two corresponding angles are equal;

or **(2) the sum of two interior angles on the same side of the transversal is equal to two right angles:**

the two straight lines are parallel.

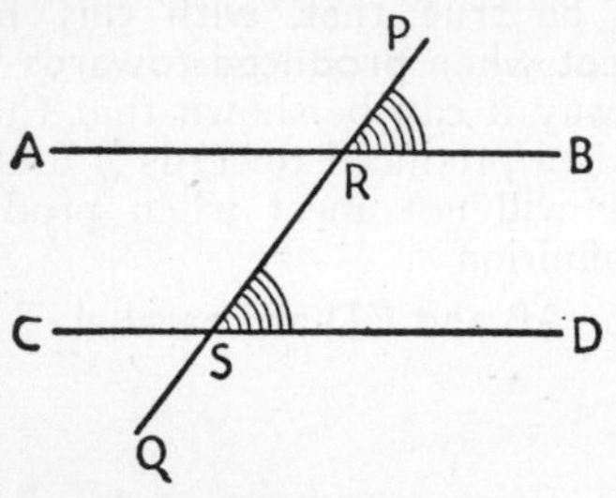

Fig. 190.

Given. The transversal PQ cuts the parallel lines AB, CD, at R and S and

(1) corr. $\angle$s PRB, RSD are equal;

or (2) $\angle BRS + \angle RSD = 2$ right $\angle$s.

To prove. AB is parallel to CD in each case.

Proofs. (1) $\angle PRB = \angle ARS$. (*Th.* 3.)

$\therefore \quad \angle ARS = \angle RSD$,

and these are alternate angles.

$\therefore$ AB is parallel to CD (*Th.* 6.)

(2) $\angle ARS + \angle BRS = 2$ rt. $\angle$s (*Th.* 1)

but $\angle RSD + \angle BRS = 2$ rt. $\angle$s (*given*)

$\therefore \quad \angle ARS + \angle BRS = \angle RSD + \angle BRS$.

Since $\angle BRS$ is common to both, on subtracting it

$$\angle ARS = \angle RSD.$$

But these are alternate angles,

$\therefore$ AB is parallel to CD (*Th.* 6)

Playfair's Axiom. *Two straight lines which intersect cannot both be parallel to the same straight line.*

Like all axioms this cannot be proved to be true, but it is self-evident and in accordance with our experience. It is necessary to assume its truth in order to prove Theorem 8.

Theorem 8

(*Converse of Theorems* 6 *and* 7)

If a straight line cuts two parallel straight lines:

(1) alternate angles are equal;

(2) corresponding angles are equal;

(3) the sum of two interior angles on the same side of the transversal is equal to two right angles.

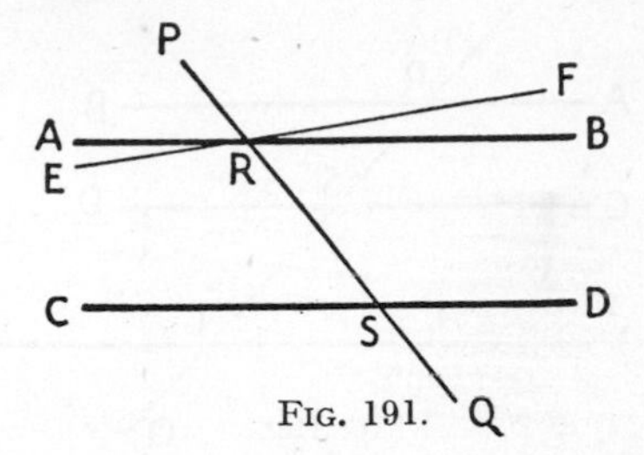

FIG. 191.

Given. AB and CD are two parallel straight lines, PQ is a transversal.

To prove. (1) alt. $\angle ARS =$ alt. $\angle RSD$.
(2) corr. $\angle PRB =$ corr. $\angle RSD$.
(3) $\angle BRS + \angle RSD = 2$ right $\angle$s.

Proof. (1) If $\angle ARS$ be not equal to $\angle RSD$, draw a straight line ERF making $\angle ERS = \angle RSD$.

But, these are alternate angles,

$\therefore$ ERF is parallel to CD (*Th.* 6)

But it is given that AB is parallel to CD, *i.e.*, two intersecting straight lines ERF and AB intersecting at R are both parallel to CD.

But this is contrary to Playfair's axiom.

$\therefore$ the assumption that $\angle ARS$ is not equal to $\angle RSD$ cannot be true,

i.e., $\angle \mathrm{ARS} = \angle \mathrm{RSD}$.

(2) Since $\angle ARS = \angle PRB$ (*Th.* 3)

and $\angle ARS = \angle RSD$ (*proved above*)

$\therefore \angle \mathrm{PRB} = \angle \mathrm{RSD}$.

(3) Since $\angle ARS + \angle BRS = 2$ rt. $\angle$s (*Th.* 1)

and $\angle ARS = \angle RSD$ (*proved above*)

$\therefore \angle \mathrm{BRS} + \angle \mathrm{RSD} = 2$ rt. $\angle$s.

Theorem 9

Straight lines which are parallel to the same straight line are parallel to one another.

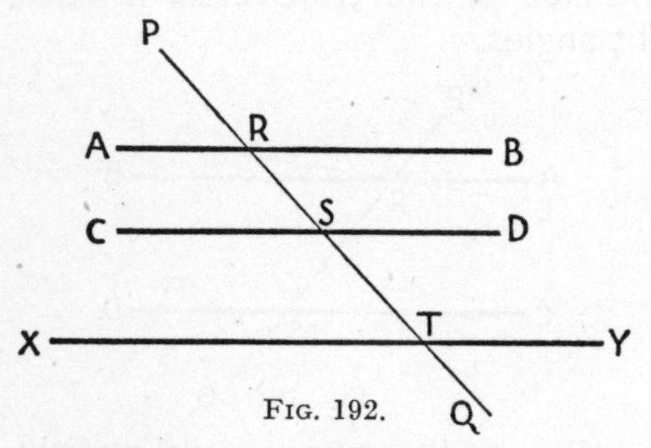

FIG. 192.

Given. AB and CD are each parallel to XY.

To prove. AB is parallel to CD.

Construction. Draw a transversal PQ cutting the straight lines AB, CD and XY in R, S, T respectively.

Proof. Since AB is parallel to XY,

$\therefore \angle PRB = \angle RTY$ (*corr.* $\angle$*s*, *Th.* 8)

Because CD is parallel to XY,

$\therefore \angle RSD = \angle RTY$ (*corr.* $\angle$*s*, *Th.* 8)

$\therefore \angle PRB = \angle RSD$.

But these are corresponding angles,

$\therefore$ AB is parallel to CD (*Th.* 7)

Exercise 25

1. AB and CD are two parallel straight lines and a transversal PQ is perpendicular to AB. Prove that it is also perpendicular to CD.

2. AB and CD are two parallels and PQ cuts them at R and S. $\angle BRS$ and $\angle RSC$ are bisected by RL and SM respectively. Prove RL is parallel to SM.

3. Two straight lines AB, CD bisect one another. Prove that AC is parallel to BD and AD is parallel to BC.

4. In a $\triangle ABC$, $\angle ABC = \angle ACB$. A straight line parallel to BC cuts AB and AC at P and Q. Prove that $\angle APQ = \angle AQP$.

5. ABC, DEF are two congruent $\triangle$s, with $AB = DE$, $BC = EF$, etc. P and Q are the mid-points of AC and DF. Prove $BP = EQ$.

6. The sides AB, AC of the $\triangle ABC$ are bisected at D and E. From these points perpendiculars are drawn to the sides and they meet in O. Join OA, OB, OC, and prove that these straight lines are equal.

SECTION 4

ANGLES OF A TRIANGLE AND POLYGON

Theorem 10

The sum of the angles of a triangle is equal to two right angles.

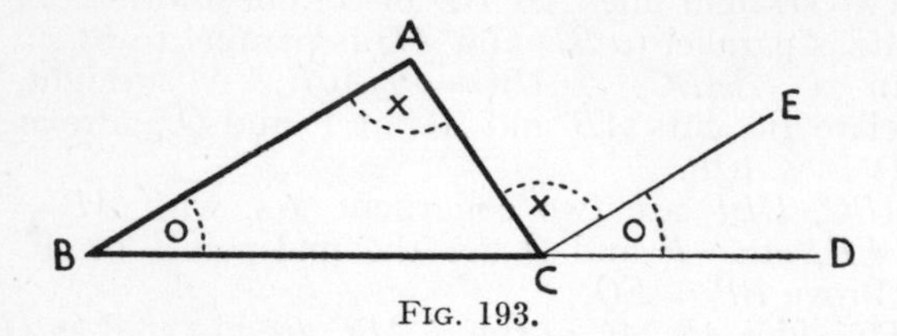

FIG. 193.

Given. ABC is any triangle.

To prove. Sum of its $\angle$s $= 2$ right $\angle$s.

Construction. Produce one of its sides, say BC, to D.
From C draw CE parallel to BA.

Proof. AB is parallel to CE, and AC cuts them.

$\therefore \quad \angle BAC = \angle ACE \quad$ (*alt.* $\angle$*s, Th.* 8)

Also AB is parallel to CE and BC cuts them.

$\therefore \quad \angle ABC = \angle ECD \quad$ (*corr.* $\angle$*s, Th.* 8)

*Adding $\quad \angle BAC + \angle ABC = \angle ACE + \angle ECD.$
Add $\angle ACB$ to both sides.

Then

$$\angle BAC + \angle ABC + \angle ACB = \angle ACE + \angle ECD + \angle ACB$$
$$= 2 \text{ right } \angle\text{s} \quad (Th.\ 1)$$

$\therefore$ sum of angles of the $\triangle = 2$ right $\angle$s.

Note.—For corollaries to this very important theorem *see* § 60 of Part I.

It may be specially noted, however, that in the step marked above with an asterisk, it is proved that

An exterior angle of a triangle is equal to the sum of the two interior opposite angles.

Theorem 11

The sum of all the interior angles of a convex polygon, together with four right angles, is equal to twice as many right angles as the figure has sides.

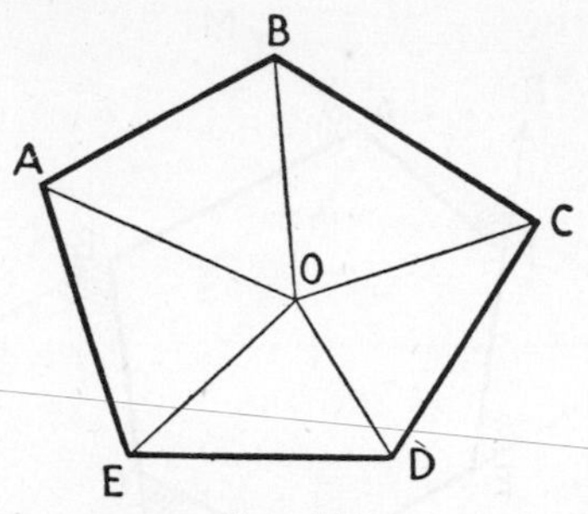

Fig. 194.

Given. Let $ABCDE$ be a polygon of n sides.

To prove. Sum of angles at A, B, C, D, E + 4 right $\angle$s = $2n$ right $\angle$s.

Construction. Let O be any point within the polygon.
Join O to the angular points A, B, C, D, E.

Proof. The polygon is composed of as many $\triangle$s as there are sides, *i.e.*, with n sides there are n $\triangle$s.

And sum of the $\angle$s of each triangle = 2 right $\angle$s. (*Th.* 10.)

$\therefore$ sum of the angles of the n $\triangle$s = 2n right $\angle$s.

But this sum is made up of

(1) the int. $\angle$s of the polygon;
(2) the angles at O;

and the angles at O = 4 right $\angle$s.

$\therefore$ all the int. $\angle$s + 4 right $\angle$s = 2n right $\angle$s.

Note.—In the case of regular polygons (§ 105, Part I) this result may be expressed algebraically as follows:

Let x° = each of the equal angles of a regular polygon of n sides.
Then $nx^\circ + 360^\circ = 180n$ or as shown in § 108, Part I.

Theorem 12

If the sides of a convex polygon are produced in the same sense, the sum of all the exterior angles so formed is equal to four right angles.

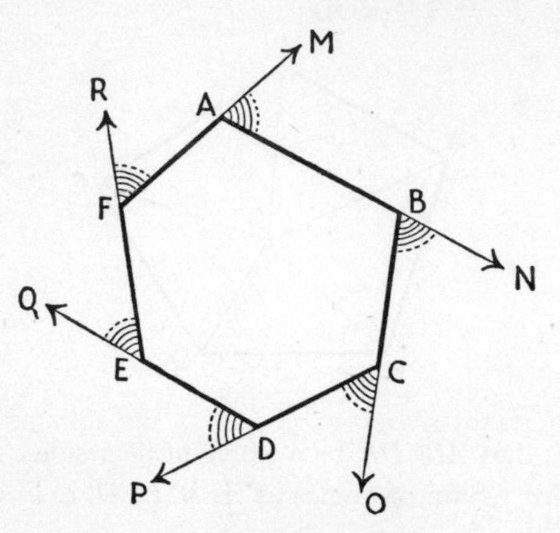

FIG. 195.

Given. ABC . . . is a convex polygon of n sides, having its sides AB, BC, CD . . . produced in the same sense (in this case, clockwise, *see* § 15) to M, N, O . . . forming exterior angles MAB, NBC, OCD . . .

To prove. Sum of ext. $\angle$s,

$$MAB + NBC + OCD + \ldots = 4 \text{ right } \angle\text{s}.$$

Proof. At each vertex A, B, C . . .

the ext. $\angle$ + int. $\angle$ = 2 right $\angle$s (*Th.* 1)

∴ for all the n vertices

sum of int. $\angle$s + sum of ext. $\angle$s = $2n$ right $\angle$s.

But

sum of int. $\angle$s + 4 right $\angle$s = $2n$ right $\angle$s (*Th.* 11)

∴ sum of ext. $\angle$s = 4 right $\angle$s.

Exercise 26

1. A straight line cuts two parallel straight lines; prove that the bisectors of two interior angles on the same side of the line are at right angles.

2. ABC is a right-angled triangle with $\angle A$ a right angle. If AD is drawn perpendicular to BC prove that $\angle DAC = \angle ABC$.

3. In a $\triangle ABC$, the side BC is produced to P and the bisector of the $\angle BAC$ meets BC in O. Prove that $\angle ABC + \angle ACP =$ twice $\angle AOP$.

4. ABC is a right-angled triangle and A is the right angle. The sides of the $\triangle$ are produced in the same sense. Prove that the sum of the exterior angles at B and C is three times the exterior angle at A.

5. From the vertex B of the $\triangle ABC$, BX is drawn perpendicular to AC, and from C, CY is drawn perpendicular to AB. Prove $\angle ABX = \angle ACY$.

6. A pentagon has one of its angles a right angle, and the remaining angles are equal. Find the number of degrees in each of them.

7. If the angles of a hexagon are all equal, prove that the opposite sides are parallel.

SECTION 5

TRIANGLES (CONGRUENT AND ISOSCELES)

Theorem 13

Two triangles are congruent if two angles and a side of one are respectively equal to two angles and a side of the other.

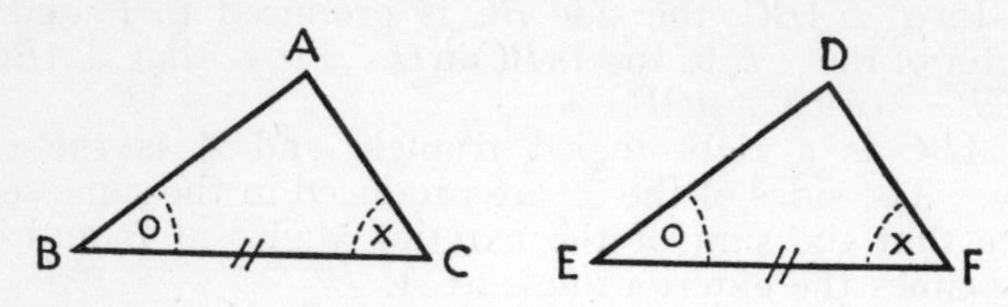

FIG. 196.

Given. ABC, DEF are $\triangle$s in which

(1) $\angle ABC = \angle DEF$.
(2) $\angle ACB = \angle DFE$.
(3) $BC = EF$.

To prove. $\triangle$s ABC, DEF are congruent.

Proof. (1) Since $\angle ABC + \angle ACB = \angle DEF + \angle DFE$. and the sum of the angles of the $\triangle$ is 2 right $\angle$s,

$\therefore$ Remaining $\angle BAC = \angle EDF$.

(2) Let the $\triangle ABC$ be placed on the $\triangle DEF$ so that BC lies along EF. Then B coincides with E and C with F.

Since $\angle ABC = \angle DEF$,

$\therefore$ BA will lie along ED, and A must lie on ED or ED produced.

Also, since $\angle ACB = \angle DFE$,

$\therefore$ CA will lie along FD, and A must lie on FD or FD produced.

$\therefore$ **A must lie on the intersection of ED and FD, *i.e.*, at D.**

Since B falls on E, C on F and A on D,

$\therefore$ $\triangle ABC$ coincides with $\triangle DEF$.

$\therefore$ **$\triangle$s ABC, DEF are congruent.**

Theorem 14

If two sides of a triangle are equal the angles opposite to these sides are equal.

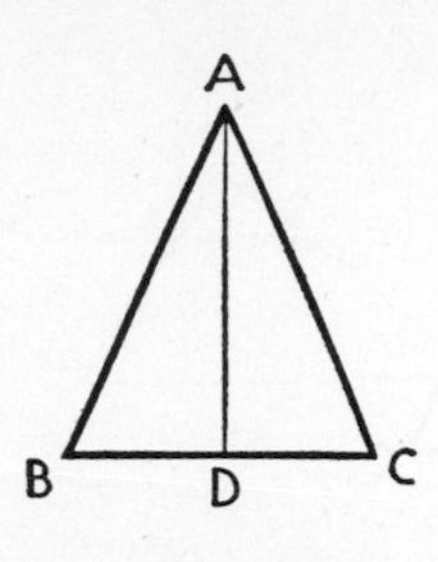

FIG. 197.

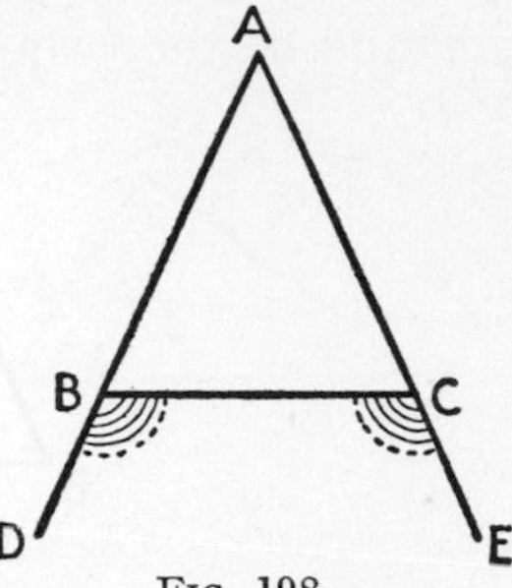

FIG. 198.

Given. In the $\triangle ABC$, $AB = AC$.

To prove. $\angle ABC = \angle ACB$.

Construction. Let AD be the straight line which bisects the $\angle BAC$, and meets BC at D.

Proof. In $\triangle$s ABD, ACD :

(1) $AB = AC$ (*given*)
(2) AD is common to both $\triangle$s
(3) $\angle BAD = \angle CAD$ (*constr.*)

$\therefore$ $\triangle$s ABD, ACD are congruent (*Th.* 4)

In particular $\angle ABC = \angle ACB$.

Corollary. If the equal sides of an isosceles triangle be produced as in Fig. 198, the exterior angles so formed are equal.

These angles are supplementary to the angles at the base which have been proved equal above.

Note.—For further corollaries, see § 62, Part I.

Theorem 15

(*Converse of Theorem* 14)

If two angles of a triangle are equal the sides opposite to these are also equal.

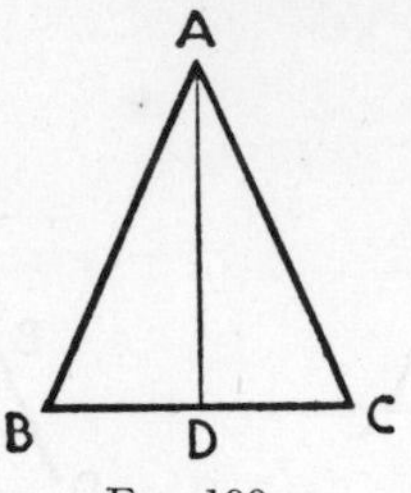

Fig. 199.

Given. ABC is a triangle in which $\angle ABC = \angle ACD$.

To prove. $AB = AC$.

Construction. Let AD be the bisector of $\angle BAC$, meeting BC in D.

Proof. In $\triangle$s ABD, ACD:

(1) $\angle ABD = \angle ACD$ (*given*)
(2) $\angle BAD = \angle CAD$ (*constr.*)
(3) AD is a side of each $\triangle$.

$\therefore$ $\triangle$s ABD, ACD are congruent (*Th.* 13)

In particular AB = AC.

Theorem 16

If in two triangles the three sides of the one are respectively equal to the three sides of the other, the triangles are congruent.

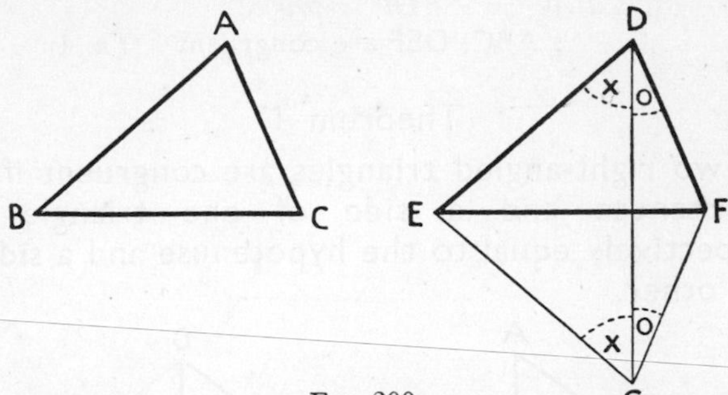

FIG. 200.

Given. ABC, DEF are $\triangle$s in which $AB = DE$, $BC = EF$, $AC = DF$.

To prove. The $\triangle$s ABC, DEF are congruent.

Proof. Let BC and EF be sides which are not the shortest in the $\triangle$s.

Apply $\triangle ABC$ to $\triangle DEF$ so that B falls on E, and BC along EF.

Since $BC = EF$, **C must coincide with F.**

Let the $\triangle ABC$ be placed so that the vertices A and D lie on opposite sides of EF, A being at G.

Join DG.

Since $EG = BA$, and $BA = ED$.

$$\therefore \quad ED = EG.$$

Similarly, $\quad FD = FG.$

$\therefore$ **$\triangle$s EGD, FGD are isosceles.**

$\therefore$ In $\triangle EGD$, $\quad \angle EGD = \angle EDG \quad$ (*Th.* 13)

Similarly, $\quad \angle FGD = \angle FDG \quad$ (*Th.* 13)

Adding $\angle EGD + \angle FGD = \angle EDG + \angle FDG$,

i.e., $\quad \angle EDF = \angle EGF.$

But $\angle BAC = \angle EGF$ (*constr.*)

$\therefore$ $\angle$BAC $=$ $\angle$EDF.

In $\triangle$s ABC, DEF:

(1) $AB = DE$ (*given*)
(2) $AC = DF$ (*given*)
(3) $\angle BAC = \angle EDF$ (*proved*)

$\therefore$ $\triangle$s ABC, DEF are congruent (*Th.* 4)

Theorem 17

Two right-angled triangles are congruent if the hypotenuse and a side of one triangle are respectively equal to the hypotenuse and a side of the other.

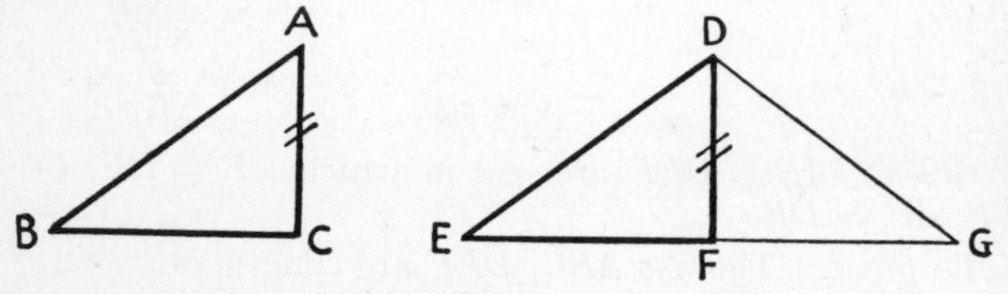

FIG. 201.

Given. $\triangle$s *ABC, DEF* are right-angled at *C* and *F*

hypotenuse *AB* = hypotenuse *DE*

and *AC* = *DF*.

To prove. $\triangle$s *ABC, DEF* are congruent.

Construction. Produce *EF* to *G* making *FG* = *BC*.
Join *DG*.

Proof. Since *EFG* is a straight line, and $\angle DFE$ is a right $\angle$.

$\therefore$ $\angle DFG$ is a right $\angle$.

In $\triangle$s ABC, DGF:

(1) $AC = DF$ (*given*)
(2) $BC = FG$ (*constr.*)
(3) $\angle ACB = \angle DFG$ (*right* $\angle$*s*)

$\therefore$ $\triangle$s ABC, DGF are congruent (*Th.* 4)

In particular, $AB = DG$, $\angle ABC = \angle DGF$.

But $AB = DE$ (*given*)

$\therefore$ $DE = DG$.

$\therefore$ $\triangle DEG$ is isosceles.

$\therefore$ $\angle DEF = \angle DGF$.

But $\angle DGF = \angle ABC$.

$\therefore$ $\angle DEF = \angle ABC$.

In $\triangle$s ABC, DEF:

(1) $AC = DF$ (*given*)

(2) $\angle ACB = \angle DFE$ (*given*)

(3) $\angle ABC = \angle DEF$ (*proved*)

$\therefore$ **$\triangle$s ABC, DEF are congruent** (*Th.* 13)

Exercise 27

1. Prove that the straight line which joins the vertex of an isosceles triangle to the mid-point of the base, (1) bisects the vertical angle, (2) is perpendicular to the base.

2. Two isosceles $\triangle$s have a common base. Prove that the straight line which joins their vertices, produced, if necessary, (1) bisects both vertical angles and (2) bisects the base at right $\angle$s.

3. Two circles intersect. Prove that the straight line joining their points of intersection is bisected at right angles by the straight line which joins their centres.

4. ABC, DEF are two acute-angled $\triangle$s, in which $AB = DE$, $AC = DF$, and the perpendicular from A to BC is equal to the perpendicular from D to EF. Prove that the $\triangle$s ABC, DEF are congruent.

5. ABC is an equilateral $\triangle$ and on its sides the equilateral $\triangle$s, ABD, BCE, ACF are constructed. Prove that DA, AF are in the same straight line, as are also DB, BE and AC, CF. Prove also that the $\triangle DEF$ which is thus formed is equilateral.

6. In any $\triangle ABC$, equilateral $\triangle$s ABD, ACE are constructed on the sides AB and AC. Prove that $BE = CD$.

7. From any point O on BD, the bisector of an angle ABC, a straight line is drawn parallel to BA or BC'. Prove that the triangle formed in each case is isosceles.

8. From a point O on AD, the bisector of an angle BAC, perpendiculars OE and OF are drawn to AB and AC respectively. Prove that $OE = OF$.

SECTION 6

INEQUALITIES

Theorem 18

If two sides of a triangle are unequal, the greater side has the greater angle opposite to it.

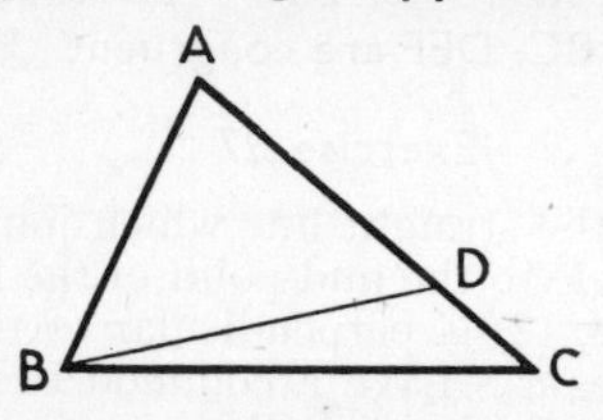

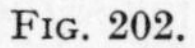
FIG. 202.

Given. In the $\triangle ABC, AC > AB$.

To prove. $\angle ABC > \angle ACB$.

Construction. From AC cut off $AD = AB$.
Join BD.

Proof. In $\triangle ABD, AB = AD$ (*constr.*)
$\therefore \angle ABD = \angle ADB$.

In $\triangle BDC$ ext. $\angle ADB >$ int. $\angle BCD$. (*Th.* 5.)
$\therefore \angle ABD > \angle BCD$.

Much more therefore is

$$\angle ABC > \angle ACB.$$

Theorem 19

(*Converse of Theorem* 18)

If two angles of a triangle are unequal, the greater angle has the greater side opposite to it.

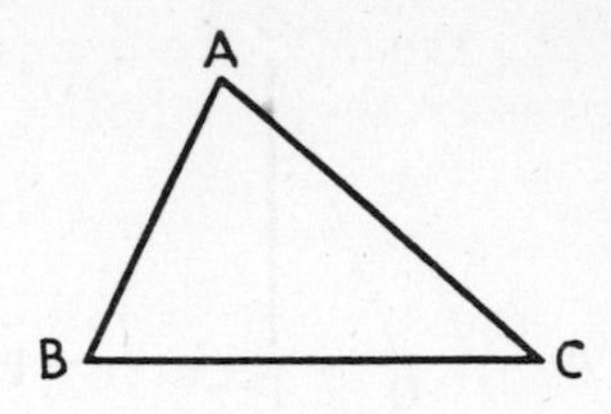

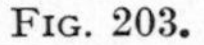
Fig. 203.

Given. In $\triangle ABC$, $\angle ABC > \angle ACB$.

To prove. $AC > AB$.

Proof. If AC is not greater than AB it must either be equal to AB or less than it.

(1) If $AC = AB$.
Then $\angle ABC = \angle ACB$. (*Th.* 14.)

But this is impossible, since $\angle ABC$ is given greater than $\angle ACB$.
$\therefore$ AC is not equal to AB.

(2) If $AC < AB$.
Then $\angle ABC < \angle ACB$. (*Th.* 18.)

But this is contrary to what is given.
$\therefore$ AC is neither equal to AB nor less than it.

$$\therefore \quad AC > AB.$$

Theorem 20

Of all the straight lines which can be drawn to a given straight line from a given point without it the perpendicular is the least.

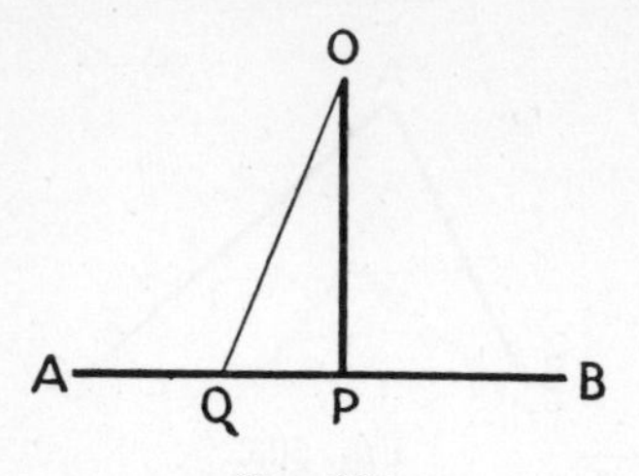

FIG. 204.

Given. O is a point lying without the straight line AB.
Draw OP perpendicular to AB.
Let OQ be any other straight line drawn from O to AB.

To prove. $OP < OQ$.

Proof. In $\triangle OPQ$, ext. $\angle OPB >$ int. $\angle OQP$.

But $\angle OPQ = \angle OPB$ (*right* $\angle s$)

$\therefore \angle OPQ > \angle OQP$.

But in the $\triangle OPQ$ the greater angle has the greater side opposite to it (*Th.* 19).

$$\therefore OQ > OP,$$

i.e., $OP < OQ$.

Theorem 21

Any two sides of a triangle are together greater than the third.

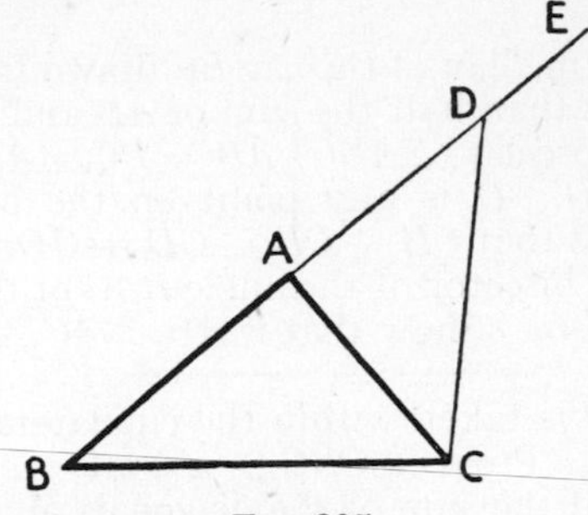

FIG. 205.

Given. ABC is any triangle.

To prove. Any two of its sides are together greater than the third; for example,

$$AB + AC > BC.$$

Construction. Produce BA to any point E.
From AE cut off $AD = AC$.
Join DC.

Proof. In $\triangle ACD$, $AD = AC$ (*constr.*)

$\therefore \quad \angle ACD = \angle ADC.$ (*Th.* 14)

But $\quad \angle BCD > ACD,$

$\therefore \quad \angle BCD > \angle BDC,$

$\therefore \quad BD > BC.$ (*Th.* 19)

But since $\quad AD = AC$ (*constr.*)

$\therefore \quad BD = BA + AD$
$= BA + AC,$

$\therefore \quad BA + AC > BC.$

Similarly any other two sides can be proved greater than the third.

Exercise 28

1. A point D is taken inside a triangle ABC, and joined to B and C. Prove $\angle BDC > \angle BAC$.

2. In the triangle ABC the bisectors of the angles ABC, ACB meet in O. If AB is less than AC, prove that OB is less than OC.

3. AD is the median of the $\triangle ABC$ drawn from A. Prove that AD is less than half the sum of AB and AC.

4. In the isosceles $\triangle ABC$, $AB = BC$. AB is produced to any point O. D is any point on the bisector of the $\angle ABC$. Prove that $CB + OB < CD + OD$.

5. AD is the bisector of the angle BAC of the $\triangle ABC$ and it meets BC in O. Show that if $AB > AC$, then $\angle AOB > \angle AOC$.

6. A point O is taken within the equilateral $\triangle ABC$ such that $OB > OC$. Prove $\angle OBC > \angle OCA$.

7. Prove that the sum of the diagonals of a quadrilateral is greater than half the sum of the sides.

8. $ABCD$ is a quadrilateral in which $AB < BC$ and $\angle BAD < \angle BCD$. Prove $AD > CD$.

SECTION 7

PARALLELOGRAMS

Definition. *A parallelogram is a quadrilateral whose opposite sides are parallel.*

Theorem 22

(1) The opposite sides and angles of a parallelogram are equal.

(2) Each diagonal bisects the parallelogram.

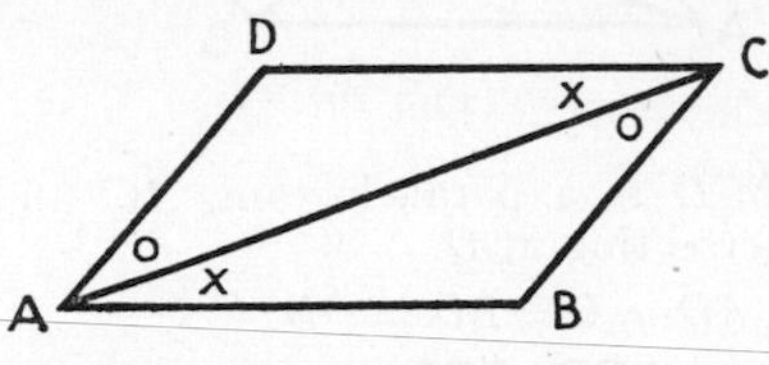

FIG. 206.

Given. $ABCD$ is a parallelogram and AC is a diagonal.

To prove. (1) Opposite sides equal, *i.e.*, $AB = DC$, $AD = BC$.

(2) Opposite $\angle$s equal, *i.e.*, $\angle ADC = \angle ABC$, $\angle BAD = \angle BCD$.

(3) The diagonal AC bisects the parallelogram.

Proof. In $\triangle$s ABC, ADC :

(1) $\angle BAC = \angle DCA$ (*alt.* $\angle$*s*, AB *is parallel to* CD, *Th.* 8)

(2) $\angle ACB = \angle DAC$ (*alt.* $\angle$*s*, AD *is parallel to* BC, *Th.* 8)

(3) AC is common to both $\triangle$s.

$\therefore$ $\triangle$s ABC, ADC are congruent (*Th.* 13)

In particular AB = DC, AD = BC.

$\angle$ADC = $\angle$ABC.

Each $\triangle$ is one half of the parallelogram, in area.

Similarly, it may be shown that $\angle BAD = \angle BCD$, and that the other diagonal, BD, bisects the parallelogram.

Theorem 23

The diagonals of a parallelogram bisect each other.

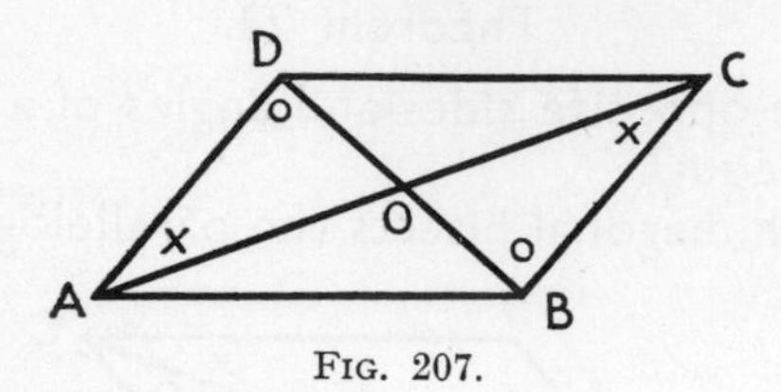

FIG. 207.

Given. $ABCD$ is a parallelogram, AC and BD are its diagonals intersecting at O.

To prove. $AO = OC$, $BO = OD$.

Proof. In $\triangle$s AOD, COB :

(1) $\angle OAD = \angle OCB$ (*alt.* $\angle$*s*, *Th.* 8)
(2) $\angle ODA = \angle OBC$ (*alt.* $\angle$*s*, *Th.* 8)
(3) $AD = BC$ (*Th.* 22)

$\therefore$ $\triangle$s AOD, COB are congruent.

In particular AO = OC,
BO = OD.

Theorem 24

(*Converse of Theorems* 22 *and* 23)

A quadrilateral is a parallelogram if

(1) its opposite sides are equal,
or **(2) its opposite angles are equal,**
or **(3) its diagonals bisect one another.**

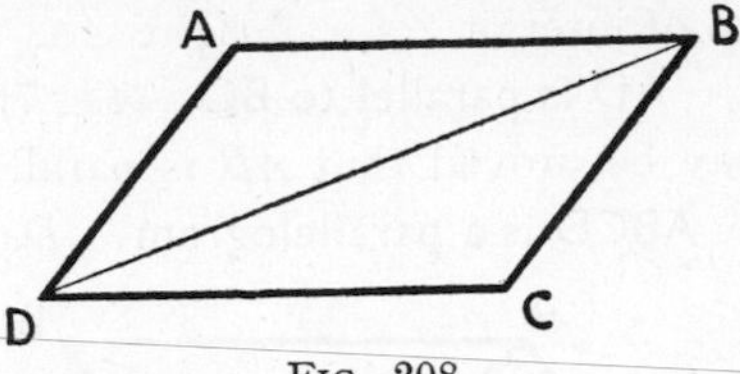

FIG. 208.

Part (1) *Given.* $ABCD$ is a quadrilateral in which

$AB = DC, AD = BC.$

To prove. $ABCD$ is a parallelogram, *i.e.*, AB is parallel to CD, AD is parallel to BC.

Construction. Joint BD.

Proof. In Δs ABD, BCD:

(1) $AB = DC$ (*given*)
(2) $AD = BC$ (*given*)
(3) BD is common.

$\therefore$ Δs ABD, BCD are congruent (*Th.* 16)

In particular $\angle ADB = \angle DBC$.

But these are alternate angles.

$\therefore$ AD is parallel to BC.

Similarly AB is parallel to CD.

$\therefore$ ABCD is a parallelogram (*Def.*)

Part (2). *Given* $\angle ADC = \angle ABC$
and $\angle DAB = \angle BCD$.

To prove. $ABCD$ is a parallelogram.

Proof. Denoting the angles of the parallelogram by $\angle A$, $\angle B$, $\angle C$, $\angle D$.

$$\angle A + \angle B + \angle C + \angle D = 4 \text{ right } \angle\text{s}.$$

But $\angle A = \angle C$ and $\angle B = \angle D$.

Substituting,

$$\therefore \quad 2\angle A + 2\angle B = 4 \text{ right } \angle\text{s}.$$
$$\therefore \quad \angle A + \angle B = 2 \text{ right } \angle\text{s}.$$

I.e., AD and BC, being cut by AB,

sum of interior $\angle$s $= 2$ right $\angle$s.

$\therefore$ AD is parallel to BC (*Th.* 7)

Similarly it may be proved that AB is parallel to DC,

$\therefore$ ABCD is a parallelogram (*Def.*)

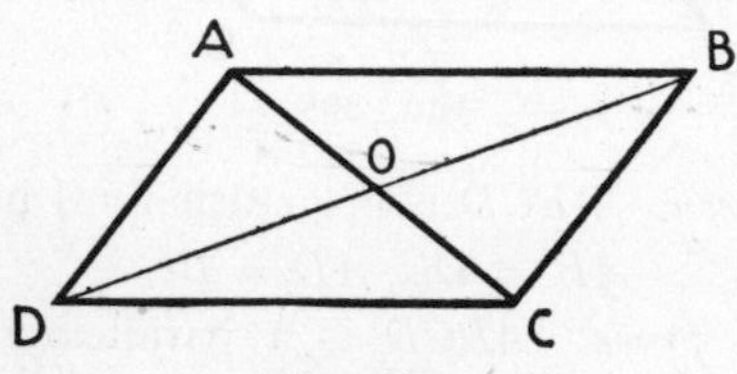

FIG. 209.

Part (3). *Given.* Diagonals AC, BD bisect one another at O, *i.e.*, $AO = OC$, $BO = OD$.

To prove. $ABCD$ is a parallelogram.

Proof. In $\triangle$s AOB, COD:

(1) $AO = OC$.
(2) $BO = OD$.
(3) $\angle AOB = \angle COD$ (*Th.* 3)

$\therefore$ $\triangle$s AOB, COD are congruent. (*Th.* 4)

In particular alt. $\angle ABO =$ alt. $\angle ODC$.

$\therefore$ AB is parallel to DC. (*Th.* 7)

Similarly it may be shown that

AD is parallel to BC.

$\therefore$ ABCD is a parallelogram. (*Def.*)

Theorem 25

The straight lines which join the ends of two equal and parallel straight lines towards the same part are themselves equal and parallel.

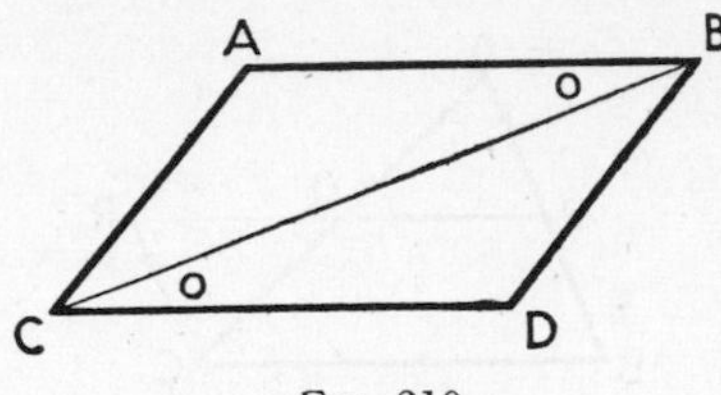

FIG. 210.

Given. AB is equal and parallel to CD.
They are joined by the straight lines AC, BD.

To prove. AC is equal and parallel to BD.

Construction. Join BC.

Proof. In $\triangle$s ABC, BCD:

(1) $AB = CD$ (*given*)
(2) BC is common.
(3) $\angle ABC = \angle BCD$ (*alt.* $\angle$*s*, *Th.* 8)

$\therefore$ $\triangle$s ABC, BCD are congruent. (*Th.* 4)

In particular $AC = BD$
and $\angle ACB = \angle DBC$.

$\therefore$ AC is parallel to BD. (*Th.* 7)

Theorem 26

A straight line drawn through the middle point of one side of a triangle and parallel to another side, bisects the third side.

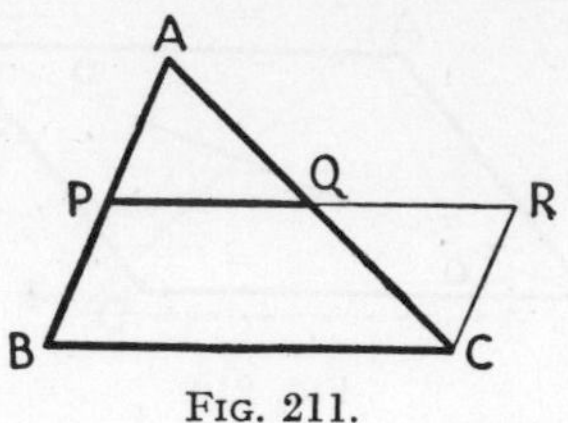

FIG. 211.

Given. From P, mid point of AB, a side of the $\triangle ABC$, PQ is drawn parallel to BC, meeting AC in Q.

To prove. AC is bisected at Q, *i.e.*,

$$AQ = QC.$$

Construction. From C draw CR, parallel to AB to meet PQ produced at R.

Proof. PR is parallel to BC (*given*)
CR is parallel to AB (*constr.*)
$\therefore$ $PRCB$ is a parallelogram (*def.*)
$\therefore$ $RC = PB = AP$ (*given*)

In $\triangle$s APQ, QRC :

(1) $AP = RC$. (*proved*)
(2) $\angle APQ = \angle QRC$ (*alt.* $\angle$*s*, AP *is parallel to* RC)
(3) $\angle PAQ = \angle QCR$ (*alt.* $\angle$*s*, AP *is parallel to* RC)

$\therefore$ $\triangle$s APQ, QRC are congruent (*Th.* 13)

In particular AQ = QC

i.e., AC is bisected at Q.

Theorem 27

The straight line joining the middle points of two sides of a triangle is parallel to the third side and equal to half of it.

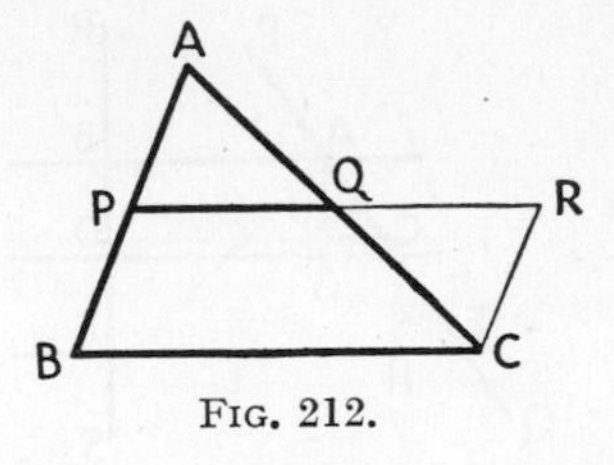

FIG. 212.

Given. P and Q are the mid points of AB, AC, sides of the $\triangle ABC$.

To prove. (1) PQ is parallel to BC.
(2) $PQ = \frac{1}{2}BC$.

Construction. Through C draw CR parallel to BP to meet PQ produced at R.

Proof. In $\triangle$s APQ, QRC:

(1) $\angle PAQ = \angle QCR$ (*alt.* $\angle$*s*, AB *is parallel to* CR)
(2) $\angle APQ = \angle QRC$ (*alt.* $\angle$*s*, AB *is parallel to* CR)
(3) $AQ = QC$ (*given*)

$\therefore$ $\triangle$s APQ, QRC are congruent. (*Th.* 13)

In particular $AP = RC$ and $PQ = QR$.
But $AP = PB$.
$\therefore$ $PB = RC$,

and PB is parallel to RC.

$\therefore$ PR and BC are equal and parallel (*Th.* 25)

But $PQ = QR$.
$\therefore$ $PQ = \frac{1}{2}PR$
$= \frac{1}{2}BC$.

$\therefore$ PQ is parallel to BC and equal to half of it.

Theorem 28

If three or more parallel straight lines make equal intercepts on any transversal they also make equal intercepts on any other transversal.

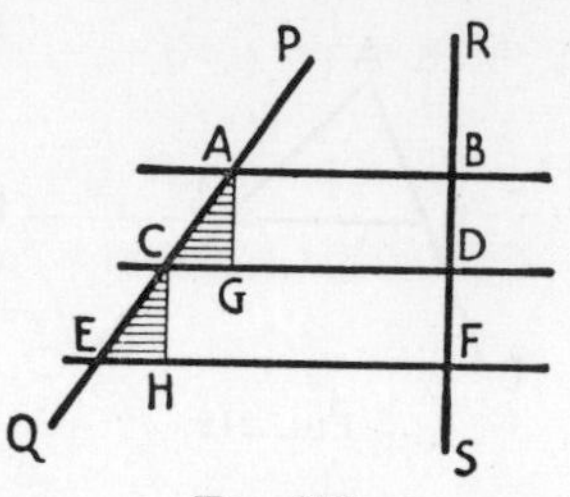

FIG. 213.

Given. AB, CD, EF are parallel straight lines.
PQ and RS are transversals.
$AC = CE$.

To prove. Intercepts on RS are equal, *i.e.*, $BD = DF$.

Construction. Draw AG parallel to BD.
Draw CH parallel to DF.

Proof. Quads. $AGDB$, $CHFD$ are parallelograms (*given and constr.*)

$\therefore$ $AG = BD$, $CH = DF$.

Since AG and CH are both parallel to RS.

$\therefore$ they are parallel to one another (*Th.* 9)

In $\triangle$s ACG, CEH :

(1) $AC = CE$ (*given*)
(2) $\angle ACG = \angle CEH$ (*corr.* $\angle$*s*, *Th.* 8)
(3) $\angle CAG = \angle ECH$ (AG *is parallel to* CH, *Th.* 8)

$\therefore$ $\triangle$s ACG, CEH are congruent. (*Th.* 13)

In particular $AG = CH$.
But $AG = BD$, and $CH = DF$.

$\therefore$ BD = DF.

Exercise 29

1. $ABCD$ is a parallelogram. From A and B perpendiculars AP, BQ are drawn to meet CD or CD produced. Prove $AP = BQ$.

2. E and F are the mid points of AB and AC, two sides of the $\triangle ABC$. P is any point on BC. AP cuts EF at Q. Prove that $AQ = PQ$.

3. E and F are the mid points of the sides AB and CD respectively of the parallelogram $ABCD$. Prove that $AECF$ is a parallelogram.

4. $ABCD$ is a parallelogram and its diagonals intersect at O. Through O a straight line is drawn cutting AB in P and CD in Q. Prove that $OP = OQ$.

5. Prove that in any quadrilateral the straight lines joining the mid points of the sides form a parallelogram.

6. $ABCD$ is a parallelogram. The bisectors of the angles A and C meet the diagonal BD in P and Q respectively. Prove that the $\triangle$s APB, CQD are congruent.

7. In the quadrilateral $ABCD$, $AB = CD$; also $\angle ABC = \angle BCD$. Prove that AD and BC are parallel.

8. A $\triangle ABC$ is right-angled at B. An equilateral $\triangle BCD$ is constructed on BC. Prove that the straight line drawn from D parallel to AB, bisects AC.

SECTION 8

AREAS

Area of a rectangle. In Part I, § 91, the rule for finding the area of a rectangle was determined. This is assumed as fundamental in the theorems which follow, as also the meaning of altitude.

Theorem 29

The area of a parallelogram is equal to that of a rectangle having the same base, and the same altitude, or between the same parallels.

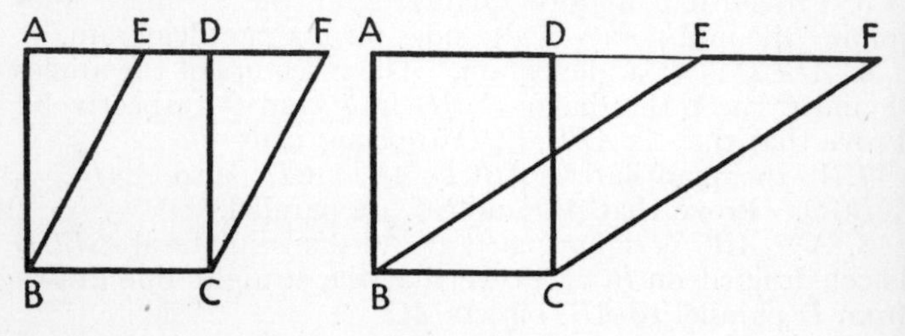

Fig. 214.

Given. The rectangle *ABCD* and the parallelogram *EBCF* have the same base *BC* and are between the same parallels *AF*, *BC*, *i.e.*, they have the same altitude which is equal to *AB* or *DC*. (There are two cases as shown by the two figures.)

To prove. Area of parallelogram *EBCF* = area of rect. *ABCD*.

Proof. In △s ABE, DCF:

(1) $\angle AEB = \angle DFC$ (*corr.* ∠*s*, BE, CF *parallel*)
(2) $\angle BAE = \angle CDF$ (*corr.* ∠*s*, AB, DC *parallel*)
(3) $AB = DC$ (*opp. sides of rectangle*)

∴ △s ABE, DCF are congruent.

Considering the whole figure *ABCF*, if the equal △s

ABE, *CDF* are subtracted *in turn* from it, the remaining figures must be equal

i.e., parallelogram EBCF = rect. ABCD.

Corollary (1). The area of a parallelogram is measured by the product of the measures of the lengths of its base and its altitude.

Corollary (2). Parallelograms on equal bases and of equal altitudes are equal in area, the area of each being measured as stated in Cor. (1).

Corollary (3). Parallelograms on the same base and between the same parallels, or having equal altitudes, are equal in area.

Theorem 30

The area of a triangle is equal to one half of the area of a rectangle on the same base and between the same parallels, or, having the same altitude.

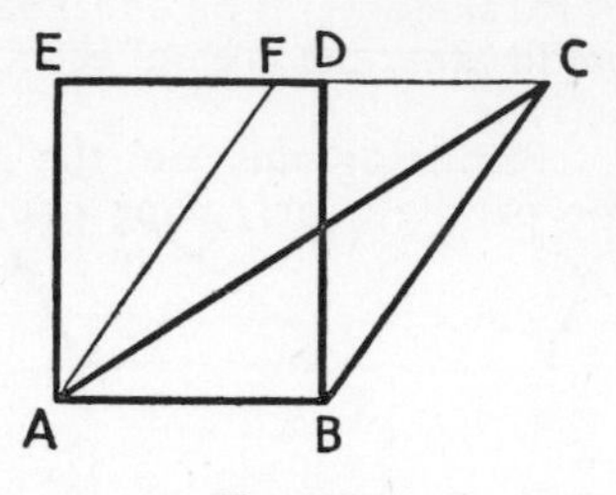

FIG. 215.

Given. The $\triangle ABC$ and rect. $ABDE$ are on the same base AB and lie between the same parallels AB and EC.

To prove. Area of $\triangle = \frac{1}{2}$ area of rectangle.

Construction. From A draw AF parallel to BC.

Proof. Since FC is parallel to AB (*given*)
and AF is parallel to BC (*constr.*)

$\therefore$ $ABCF$ is a parallelogram and AC is a diagonal.

$\therefore$ $\triangle ABC = \frac{1}{2}$ parallelogram $ABCF$ (*Th.* 22)
But parallelogram $ABCF$ = rect. $ABDE$ in area (*Th.* 29)
$\therefore$ $\triangle ABC = \frac{1}{2}$ rect. ABDE.

Corollary 1. The area of a triangle is equal to one half the product of the measures of its base and altitude,

or Area $= \frac{1}{2}$ (base $\times$ altitude).

Corollary 2. Triangles on the same base, or on equal bases and having the same altitude are equal in area.

Corollary 3. If triangles with the same area have the same or equal bases, their altitudes are equal.

Theorem 31

Triangles of equal area, which stand on equal bases, in the same straight line, and on the same side of it, are between the same parallels.

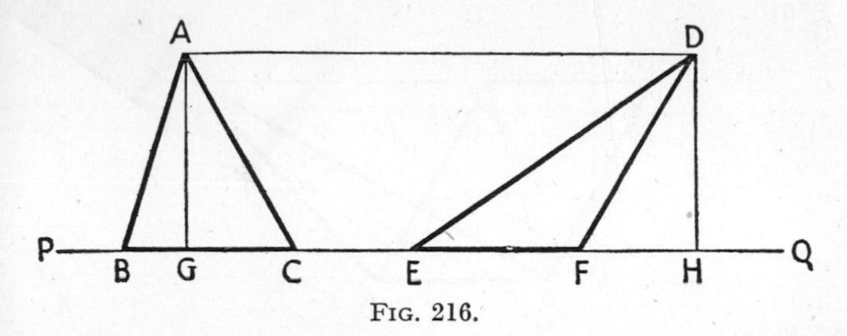

FIG. 216.

Given. ABC, DEF are $\triangle$s of the same area, standing on equal bases BC, EF, in the same straight line, PQ.
Join AD.

To prove. AD is parallel to PQ.

Construction. Draw AG, DH perpendicular to PQ.

Proof. Since $\triangle$s ABC, DEF have equal areas and stand on equal bases.

$\therefore$ their altitudes are equal (*Th.* 30, *Cor.* 3)

i.e., $AG = DH$.

Since AG and DH are perpendicular to PQ.

$\therefore$ AG is parallel to DH,

i.e., AG and DH are equal and parallel.

$\therefore$ **AD and PQ are parallel** (*Th.* 25)

Theorem 32

If a triangle and a parallelogram stand on the same base and are between the same parallels, the area of the triangle is half that of the parallelogram.

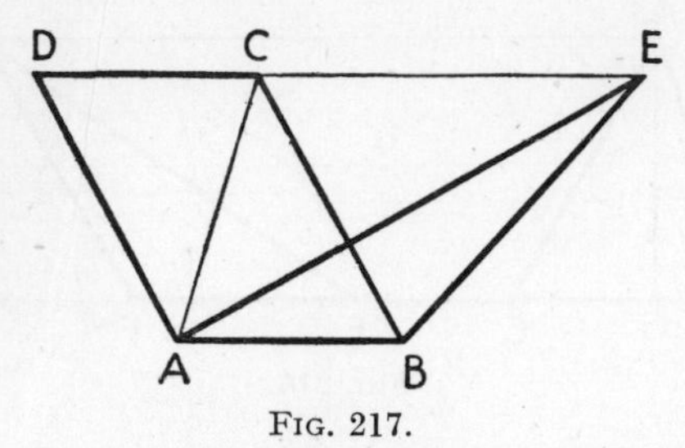

FIG. 217.

Given. Parallelogram $ABCD$ and $\triangle ABE$ stand on the same base AB and are between the same parallels AB and DE.

To prove. Area of $\triangle ABE$ = half the area of parallelogram $ABCD$.

Construction. Join AC.

Proof. AB and DE are parallels.

$\therefore$ area of $\triangle ACB$ = area of $\triangle AEB$ (*Th.* 30, *Cor.* 2)

But $\triangle ACB = \frac{1}{2}$ parallelogram $ABCD$. (*Th.* 22)

$\therefore$ $\triangle ABE = \frac{1}{2}$ parallelogram ABCD.

SECTION 9

RIGHT-ANGLED TRIANGLES

Theorem 33

(Theorem of Pythagoras)

The area of the square on the hypotenuse of a right-angled triangle is equal to the sum of the areas of the squares on the two sides which contain the right angle (*see* § 99).

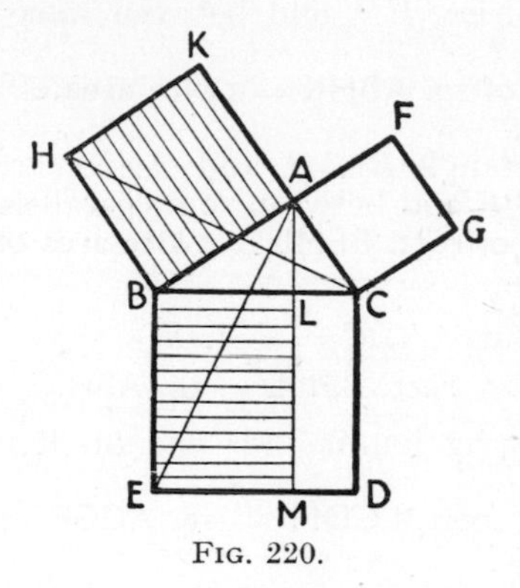

Fig. 220.

Given. ABC is a right-angled triangle, $\angle A$ the right angle.

To prove. Sq. on BC = sq. on AB + sq. on AC.

Construction. On BC, AC and AB construct the squares $BCDE$, $ACGF$, $ABHK$.

Draw ALM parallel to BE and CD, meeting BC in L, and DE in M.

Join AE and CH.

Proof.

Step 1. $\angle$s BAC and BAK are right $\angle$s.

$\therefore$ KA and AC are in the same straight line.

Similarly BA and AF are in the same straight line.

Step 2. $\angle HBA = \angle EBC$ (*right* $\angle$*s*)

Add $\angle ABC$ to each.

$$\therefore \quad \angle HBC = \angle ABE.$$

Step 3. In $\triangle$s ABE, HBC :

(1) $\angle ABE = \angle HBC$ (*proved above*)
(2) $AB = HB$ (*sides of square*)
(3) $BE = BC$ (*sides of square*)

$\therefore$ $\triangle$s ABE, HBC are congruent (*Th.* 4)

Step 4. The square $ABHK$ and $\triangle HBC$ are on the same base, HB, and between same parallels, HB and KC.

$\therefore$ area of sq. ABHK = twice area of $\triangle$HBC (*Th.* 32)

Also rectangle $BEML$ and $\triangle ABE$ are on the same base BE and between same parallels BE and AM.

$\therefore$ area of rect. BEML = twice area of $\triangle$ABE (*Th.* 32)

Step 5. But $\triangle HBC = \triangle ABE$.

$$\therefore \text{rect. BEML} = \text{sq. ABHK}.$$

Similarly by joining AD and BG it may be shown that

$$\text{rect. LCDM} = \text{sq. ACGF}.$$

Step 6.

$\therefore$ rect. $BEML$ + rect. $LCDM$ = sq. $ABHK$ + sq. $ACGF$,
i.e., sq. $BCDE$ = sq. $ABHK$ + sq. $ACGF$
or sq. on BC = sq. on AB + sq. on AC.

Theorem 34

(*Converse of Theorem* 33)

If the square on one side of a triangle is equal to the sum of the squares on the other two sides, then the angle contained by these two sides is a right angle.

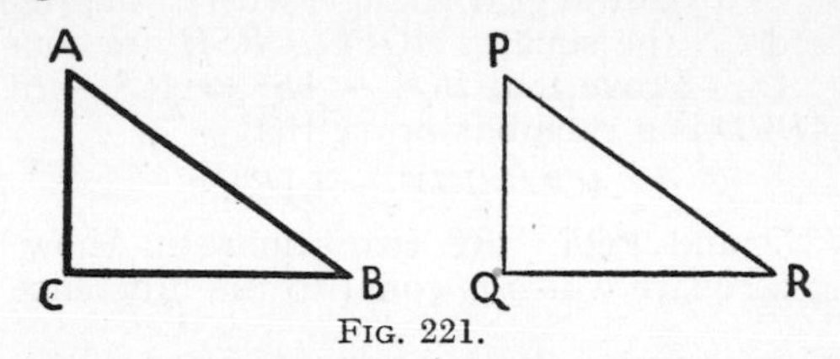

FIG. 221.

Given. ABC is a triangle in which

$$AB^2 = AC^2 + CB^2.$$

To prove. $\angle ACB$ is a right angle.

Construction. Construct a $\triangle PQR$ such that $PQ = AC$, $\angle PQR$ is a right angle, $QR = BC$.

Proof. Since $\angle PQR$ is a right angle

$PR^2 = PQ^2 + QR^2$ (*Th.* 33)

Since $PQ = AC$, and $QR = BC$

$\therefore$ $PR^2 = AC^2 + BC^2$,

but $AB^2 = AC^2 + BC^2$ (*given*)

$\therefore$ $PR^2 = AB^2$

and $PR = AB$.

In $\triangle$s ABC, PQR:

(1) $PR = AB$ (*proved*)
(2) $PQ = AC$ (*constr.*)
(3) $QR = BC$ (*constr.*)

$\therefore$ $\triangle$s ABC, PQR are congruent (*Th.* 16)

In particular $\angle ACB = \angle PQR$.

But $\angle PQR$ is a right angle (*constr.*)

$\therefore$ **$\angle$ACB is a right angle.**

Exercise 31

1. Prove that the sum of the squares on the sides of a rectangle is equal to the sum of the squares on its diagonals.

2. ABC is any triangle and AD is the perpendicular drawn from A to BC.

Prove that $AB^2 - BD^2 = AC^2 - DC^2$.

3. ABC is a right-angled triangle with C the right angle. On AC and CB the squares $AQPC$, $CRSB$ are constructed. Join BQ, AS. Prove that $BQ^2 - AS^2 = AC^2 - BC^2$.

4. If $ABCD$ is a rhombus prove that

$$AC^2 + BD^2 = 4AB^2.$$

5. $ABCD$ and $PQRS$ are two squares. Show how to construct a square which is equal to the difference of their areas.

6. B is the right angle in a right-angled triangle ABC. Any point P is taken on BC. Prove that

$$AP^2 + BC^2 = BP^2 + AC^2.$$

7. On a straight line BC two equilateral $\triangle$s ABC, DBC are constructed on opposite sides of BC. Join AD. Prove that the square on AD is equal to three times the square on BC.

8. Any point O is taken within a rectangle $ABCD$. Prove that

$$OA^2 + OC^2 = OB^2 + OD^2.$$

SECTION 10

EXTENSIONS OF THEOREM OF PYTHAGORAS

Theorem 35

Note.—Before proceeding to the two following theorems the student is advised to revise Part I, § 152. The proofs given below are algebraical.

In an obtuse-angled triangle the square on the side opposite to the *obtuse* angle is equal to the sum of the squares on the sides containing the obtuse angle *plus* twice the rectangle contained by either of these sides and the projection on it of the other.

Given. ABC is a triangle with an obtuse angle at C.

AD is the perpendicular from A to BC produced.

Let $AD = h$, and $CD = p$.

Then p is the projection of AC upon BC (§ 30)

Let a, b, c, in units of length, represent the sides of the triangle (§ 44)

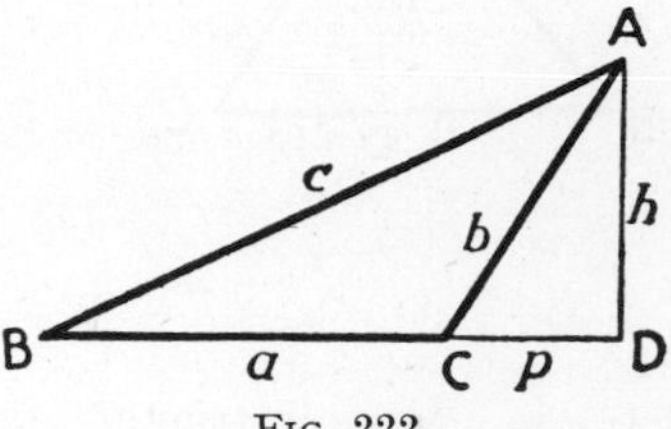

FIG. 222.

To prove. $AB^2 = AC^2 + BC^2 + 2BC \cdot CD$

or $c^2 = a^2 + b^2 + 2ap$.

Proof. $BD = BC + CD = a + p$.

Since ABD is a right-angled triangle

$$AB^2 = BD^2 + AD^2 \quad (Th.\ 33)$$

Substituting $c^2 = (a + p)^2 + h^2$.

By algebra $(a + p)^2 = a^2 + 2ap + p^2$ (*See also Appendix A*)

$$\therefore \quad c^2 = a^2 + 2ap + p^2 + h^2.$$

But $AC^2 = CD^2 + AD^2$ (*Th.* 33)

or $b^2 = p^2 + h^2$.

Substituting $c^2 = a^2 + 2ap + b^2$

or $c^2 = a^2 + b^2 + 2ap$.

Expressing this in geometric form

$$AB^2 = AC^2 + BC^2 + 2BC \cdot CD.$$

Theorem 36

In any triangle the square on the side opposite to an *acute* angle is equal to the sum of the squares on the sides containing that acute angle *less* twice the rectangle contained by one of those sides and the projection on it of the other.

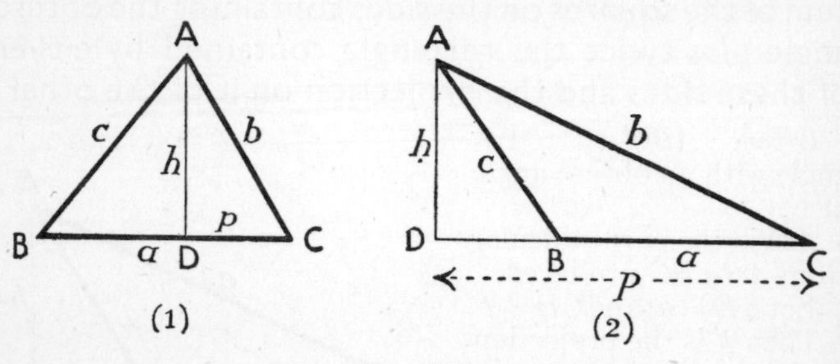

FIG. 223.

Given. In any triangle there are at least two acute angles; consequently there are two cases (shown in Fig. 223 (1) and (2)):

(1) when all the angles are acute;
(2) when one angle is obtuse.

Draw AD (denoted by h) perpendicular to BC or BC produced in (2).

Then CD (denoted by p) is the projection of AC on BC, or BC produced.

In Fig. 223 (1) $BD = a - p$; in Fig. 223 (2) $BD = p - a$.

To prove. $AB^2 = AC^2 + BC^2 - 2BC \cdot CD$

or $$c^2 = a^2 + b^2 - 2ap.$$

Proof. Considering the right-angled $\triangle ABD$.

$$AB^2 = BD^2 + AD^2. \quad (Th.\ 33)$$

$\therefore$ in (1) $$c^2 = (a - p)^2 + h^2$$
In (2) $$c^2 = (p - a)^2 + h^2.$$

From algebra $(a - p)^2 = (p - a)^2$
$= a^2 + p^2 - 2ap$ (*See Appendix A*)

$\therefore$ in (1) and (2) $c^2 = a^2 - 2ap + p^2 + h^2$.

But from $\triangle ADC$

$$p^2 + h^2 = b^2 \qquad (Th.\ 33)$$

$\therefore$ on substitution $c^2 = a^2 + b^2 - 2ap$

or in geometric form $AB^2 = AC^2 + BC^2 - 2 . BC . CD$.

Exercise 32

1. Prove that the sum of the squares on two sides of a triangle is equal to twice the sum of the square on half the third side and twice the square on the median which bisects that side. (*Apollonius' Theorem.*)

(*Hint.* Drop a perpendicular to the third side from the opposite vertex and use Theorems 35 and 36.)

2. ABC is an isosceles triangle in which $AB = AC$. CD is the perpendicular drawn from C to the opposite side. Prove that

$$BC^2 = 2AB . BD.$$

3. ABC is an acute-angled triangle and BE, CF are altitudes. Prove that

$$AF . AB = AE . AC.$$

4. In the parallelogram $ABCD$, prove that

$$AC^2 + BD^2 = 2AB^2 + 2BC^2.$$

5. In a $\triangle ABC$, D is the mid point of BC. Find the length of the median AD when $AB = 4$ in., $BC = 5$ in., and $AC = 6$ in. (Ans. 4·38 in., appr.)

(*Hint.* Use the Theorem of Apollonius, mentioned above.)

SECTION 11

CHORDS OF CIRCLES

Theorem 37

(1) The straight line which joins the centre of a circle to the middle point of a chord (which is not a diameter) is perpendicular to the chord.

Conversely:

(2) The straight line drawn from the centre of a circle perpendicular to a chord, bisects the chord.

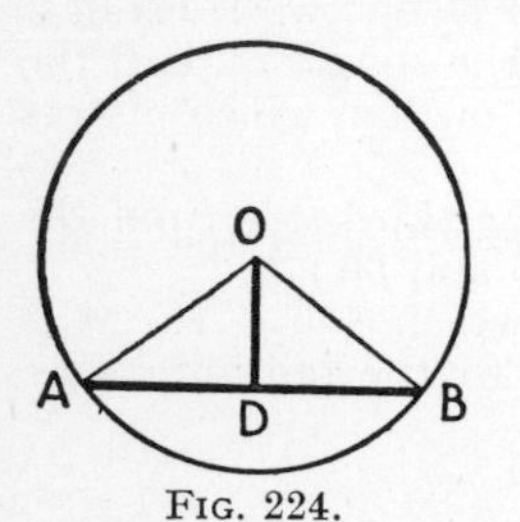

FIG. 224.

(1) *Given.* O is the centre and AB a chord of a circle.

O is joined to D the centre of AB.

To prove. OD is perpendicular to AB.

Construction. Join OA, OB.

Proof. In $\triangle$s OAD, OBD:

(1) $OA = OB$ (*radii*)
(2) $AD = DB$ (*given*)
(3) OD is common.

$\therefore$ $\triangle$s OAD, OBD are congruent (*Th.* 16)

In particular $\angle ODA = \angle ODB$.

$\therefore$ OD is perpendicular to AB.

Converse:

(2) *Given.* OD is perpendicular to AB.

To prove. AB is bisected at D, *i.e.*, $AD = DB$.

Proof. In $\triangle$s OAD, OBD:

(1) $OA = OB$ (*radii*)
(2) OD is common.
(3) $\angle$s ODA, ODB are right $\angle$s.

$\therefore$ $\triangle$s OAD, OBD are congruent. (*Th.* 17)

In particular AD = DB.

Corollary. *The perpendicular bisector of a chord passes through the centre of the circle.*

Theorem 38

One circle and one only can be drawn through three points not in the same straight line.

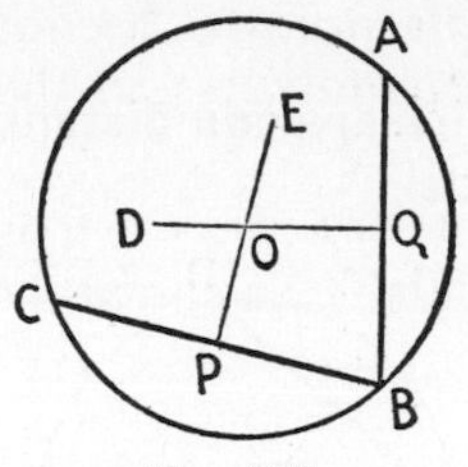

FIG. 225.

Given. A, B and C are three points not in the same straight line.

To prove. One circle and one only can pass through A, B and C.

Construction. Join AB and BC.

Draw the perpendicular bisectors of these straight lines, *i.e.*, PE and QD.

Since A, B, C are not in the same straight line, the perpendicular bisectors of AB and BC are not parallel and therefore must meet.

Let them meet at O.

Proof. Since PE is the perpendicular bisector of BC, all points equidistant from B and C lie on it. (*Proof as in* § 70.)

Similarly all points equidistant from A and B lie on the perpendicular bisector QD.

O is a point, and the only point in which these perpendicular bisectors intersect.

I.e., O is a point, and the only point which is equidistant from A, B and C.

$\therefore$ O is the centre of a circle, radius OB, which will pass through the points A, B and C. Also there is no other circle which passes through these points.

Theorem 39

(1) Equal chords of a circle are equidistant from the centre.

Conversely :

(2) Chords which are equidistant from the centre of a circle are equal.

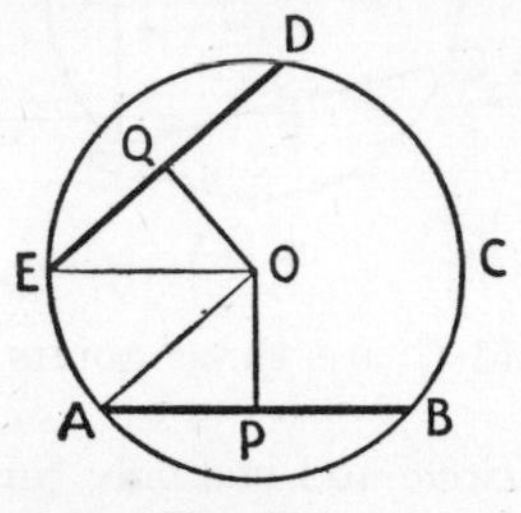

Fig. 226.

(1) *Given.* AB, DE are equal chords of a circle ABC, centre O.

OP, OQ are perpendiculars on the chords from O and therefore are the distances of the chords from O.

To prove. $OP = OQ$.

Construction. Join OA, OE.

Proof. $EQ = \frac{1}{2}ED$ and $AP = \frac{1}{2}AB$ (*Th.* 37)
and $AB = DE$ (*given*)
$\therefore$ $EQ = AP$.

In $\triangle$s OAP, OEQ:

(1) $OE = OA$ (*radii*)
(2) $EQ = AP$ (*proved*)
(3) $\angle OQE = \angle OPA$ (*right* $\angle$*s*)

$\therefore$ $\triangle$s OAP, OEQ are congruent (*Th.* 17)

In particular OP = OQ,

i.e., the chords are equidistant from the centre.

Converse :

(2) *Given.* $OP = OQ$.

To prove. $AB = DE$.

Proof. In $\triangle$s OAP, OEQ:

(1) $OE = OA$ (*radii*)
(2) $OP = OQ$ (*given*)
(3) $\angle OQE = \angle OPA$ (*right* $\angle$*s*)

$\therefore$ $\triangle$s OAP, OEQ are congruent. (*Th.* 17)

$\therefore$ $EQ = AP$.

But $ED = 2EQ$ and $AB = 2AP$.

$\therefore$ $AB = ED$.

Exercise 33

1. OA and OB are two chords of a circle which make equal angles with the straight line joining O to the centre. Prove that the chords are equal.

2. Show how to construct in a given circle a chord AB which passes through a given point O within the circle and is bisected at the point.

3. From a point A on the circumference of a circle equal chords AB and AC are drawn. If O be the centre of the circle prove $\angle OAB = \angle OAC$.

4. Two circles intersect at P and Q. Prove that the straight line joining the centres of the circles bisects the common chord PQ at right angles.

5. P is a point on a chord AB of a circle. Show how to draw through P a chord equal to AB.

6. Two circles intersect at P and Q. Through these points parallel straight lines APC, BQD are drawn to meet the circles in A, B, C, D. Prove that $AC = BD$.

7. Two concentric circles are cut by a chord $ABCD$ which intersects the outer circle in A and D and the inner in B and C. Prove that $AB = CD$.

SECTION 12

ANGLE PROPERTIES OF A CIRCLE

Theorem 40

The angle which an arc of a circle subtends at the centre is twice that which it subtends at any point on the remaining part of the circumference of the arc.

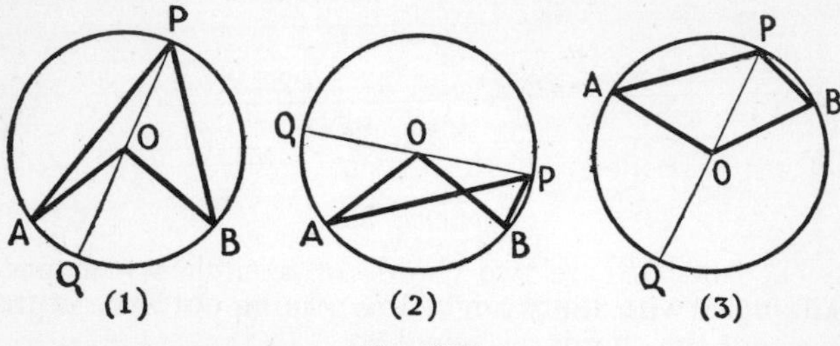

FIG. 227.

Given. AB is an arc of the circle ABP.

$\angle AOB$ is the angle subtended by the arc at the centre O.

$\angle APB$ is the angle subtended by the arc at P any point on the remaining part of the circumference.

To prove. $\angle AOB =$ twice $\angle APB$.

There are three cases :

(1) When centre O lies within the $\angle APB$.

(2) When centre O lies without the $\angle APB$.

(3) When $\angle AOB$ is a reflex angle and $\angle APB$ is obtuse.

Construction. Join PO and produce it to meet the circumference at Q.

Proof.

Case 1. In $\triangle OAP$, $OA = OP$; $\therefore \angle OAP = \angle OPA$.

But ext. $\angle AOQ = \angle OAP + \angle OPA$ (*Th.* 10, *note*)

$\therefore \angle AOQ =$ twice $\angle OPA$ . . (I)

Similarly from $\triangle OPB$,

$$\angle BOQ = \text{twice } \angle OPB \quad . \quad . \quad \text{(II)}$$

Adding (I) and (II).

$$\angle AOQ + \angle BOQ = \text{twice } (\angle OPA + \angle OPB).$$

$$\therefore \quad \angle AOB = \text{twice } \angle APB.$$

Case 2. With the same reasoning as above, but subtracting I from II.

$$\angle BOQ - \angle AOQ = \text{twice } \angle OPB - \text{twice } \angle OPA.$$

$$\therefore \quad \angle AOB = \text{twice } \angle APB.$$

Case 3. As before.

$$\angle AOQ = \text{twice } \angle APQ.$$

$$\angle BOQ = \text{twice } \angle BPQ.$$

Adding $\quad$ Reflex $\angle AOB = \text{twice } \angle APB.$

Theorem 41

Angles in the same segment of a circle are equal.

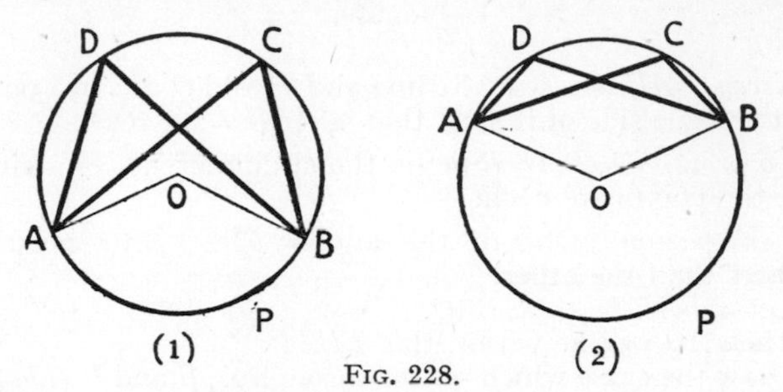

FIG. 228.

Given. The angles may be acute as in (1) or obtuse as in (2). The proof below applies to both cases.

In the circle $ABCD$, centre O, the arc APB subtends any two angles in the segment, viz., ACB, ADB.

To prove. $\quad \angle ACB = \angle ADB.$

Construction. Join OA, OB.

Proof. $\quad \angle AOB = \text{twice } \angle ADB \quad$ (*Th.* 40)

Also $\quad \angle AOB = \text{twice } \angle ACB \quad$ (*Th.* 40)

$$\therefore \quad \angle ADB = \angle ACB.$$

Theorem 42

(*Converse of Theorem* 41)

If a straight line subtends equal angles at two points on the same side of it, then these two points and the points at the extremities of the line lie on a circle.

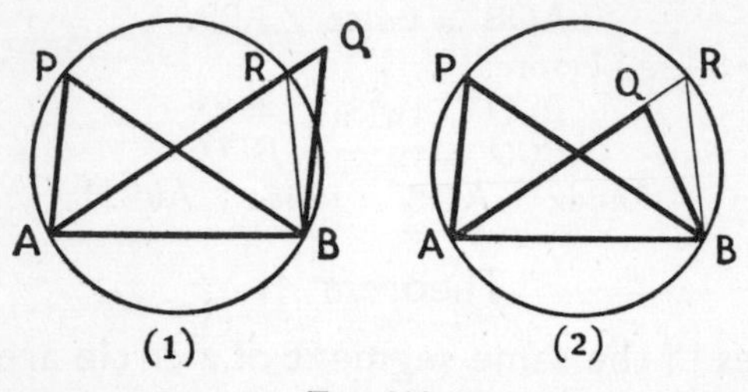

FIG. 229.

Given. AB is a straight line and P and Q are two points on the same side of it such that $\angle APB = \angle AQB$.

To prove. A, B, Q, P lie on the circumference of a circle, *i.e.*, the points are cyclic.

Construction. One of the angles BAP, BAQ must be greater than the other.

Let $\angle BAP$ be $> \angle BAQ$.

Then AQ will lie within the $\angle BAP$.

Draw the circle which passes through A, B and P (*Th.* 38).

Then, since AQ lies within the angle BAP, the circle must cut AQ (Fig. 229 (1)) or AQ produced (Fig. 229 (2)) at some point R.

Join BR.

Proof. Because A, P, R, B are cyclic, in either case and $\angle$s APB, ARB are in the same segment.

$\therefore \quad \angle APB = \angle ARB \quad$ (*Th.* 41)

But $\quad \angle APB = \angle AQB \quad$ (*given*)

$\therefore \quad \angle ARB = \angle AQB.$

This is impossible since one of the angles is the exterior and the other the interior angle of the $\triangle BRQ$.

$\therefore$ the assumption that the circle does not pass through Q has led to an absurdity and cannot be true.

$\therefore$ **A, B, P and Q lie on the circumference of a circle.**

Theorem 43

The angle in a semi-circle is a right angle.

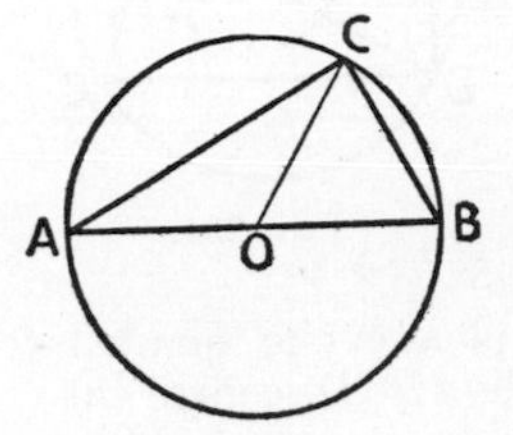

FIG. 230.

Given. AB is a diameter of the circle ACB, O is the centre and ACB any angle in the semi-circle.

To prove. $\angle ACB$ is a right angle.

Construction. Join OC.

Proof. $OA = OC$ (*radii of circle, ACB*)

$\therefore \angle OAC = \angle OCA$ (*Th.* 14)

Also $OB = OC$

$\therefore \angle OBC = \angle OCB$

$\therefore \angle OAC + \angle OBC = \angle OCA + \angle OCB$

$= \angle ACB.$

But, since the angles of $\triangle ACB$ equal 2 right $\angle$s.

$\therefore \angle ACB =$ one-half of two right $\angle$s.

I.e., **$\angle$ACB is a right $\angle$.**

Note.—See also Part I, § 135.

Theorem 44

The sum of the opposite angles of a quadrilateral inscribed in a circle is equal to two right angles, *i.e.*, the opposite angles are supplementary.

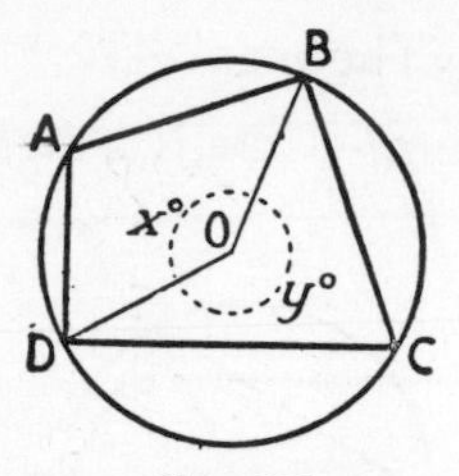

FIG. 231.

Given. $ABCD$ is a cyclic quadrilateral, *i.e.*, it is inscribed in the circle $ABCD$, centre O.

To prove. $\angle BCD + \angle BAD = 2$ right $\angle$s.
and $\angle ABC + \angle ADC = 2$ right $\angle$s.

Construction. Join O to B and D.

Proof. $\angle BOD(x^\circ) =$ twice $\angle BCD$ (*Th.* 40)
reflex $\angle BOD(y^\circ) =$ twice $\angle BAD$ (*Th.* 40)

$\therefore$ $\angle BOD +$ reflex $\angle BOD =$ twice $(\angle BCD + \angle BAD)$.

But

$\angle BOD +$ reflex $\angle BOD = 4$ right $\angle$s.
$\therefore$ $\angle BCD + BAD = 2$ right $\angle$s.

Similarly by joining OA, OC it may be proved that

$\angle ABC + \angle ADC = 2$ right $\angle$s.

Corollary. If a side of cyclic quadrilateral be produced, the exterior angle so formed is equal to the interior opposite angle.

Theorem 45

(*Converse of Theorem* 44)

If the sum of a pair of opposite angles of a quadrilateral is equal to two right angles, *i.e.*, the angles are supplementary, the quadrilateral is cyclic.

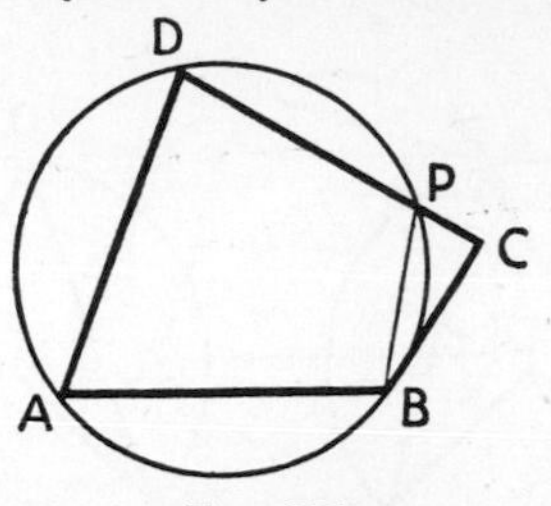

FIG. 232.

Given. $ABCD$ is a quadrilateral in which a pair of opposite $\angle$s, BAD, BCD is equal to two right angles.

To prove. The quadrilateral $ABCD$ is cyclic.

Proof. A circle can be described to pass through the three points B, A, D (*Th.* 38).

If this does not pass through C the circumference must cut DC, or DC produced at some point P.

Join BP.

$ABPD$ is a cyclic quadrilateral.

$\therefore$ $\angle BPD$ is supplementary to $\angle BAD$ (*Th.* 45)

But $\angle BCP$ is supplementary to $\angle BAD$ (*given*)

$$\therefore \quad \angle BCP = \angle BPD,$$

i.e., the exterior angle of the $\triangle BCP$ is equal to the interior opposite angle.

But this is impossible (*Th.* 5)

$\therefore$ the assumption that the circle which passes through B, A, D does not pass through C has led to an absurdity.

$\therefore$ the circle which passes through B, A, D must pass through C,

i.e., **quad. ABCD is cyclic.**

Note.—If the circle cuts DC produced, the fig. will be as in Fig. 229 (2) and the proof is similar to that above.

Theorem 46

In equal circles, or in the same circle, if two arcs subtend equal angles

(1) at the centre,

or **(2) at the circumferences,**

they are equal.

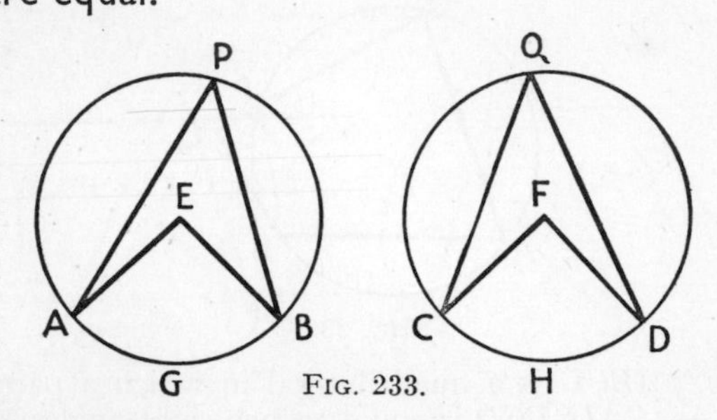

Fig. 233.

Given. Two circles of equal radii *APBG*, *CQDH*, centres *E* and *F* having

(1) equal angles *AEB*, *CFD* at the centre

or (2) equal angles *APB*, *CQD* at the circumference.

To prove. The arcs on which these angles stand, viz., *AGB*, *CHD* are equal.

Proof. (1) Apply the circle *APB* to the circle *CQD* so that the centre *E* falls on the centre *F* and *AE* along *CF*.

Since $\angle AEB = \angle CFD$, *EB* will lie along *FD*. Also, the radii are equal.

$\therefore$ A must fall on C and B on D.

$\therefore$ the arc *AGB* must coincide with the arc *CHD*,

i.e., arc AGB = arc CHD.

(2) When angles at the circumference *APB*, *CQD* are given equal.

Then, since angles at the centre are double those at the circumference.

$$\therefore \quad \angle AEB = \angle CFD.$$

$\therefore$ by the first part of the proof

arc AGB = arc CHD.

Theorem 47

(*Converse of Theorem* 46)

In equal circles, or the same circle, if two arcs are equal; they subtend equal angles at the centre and at the circumference.

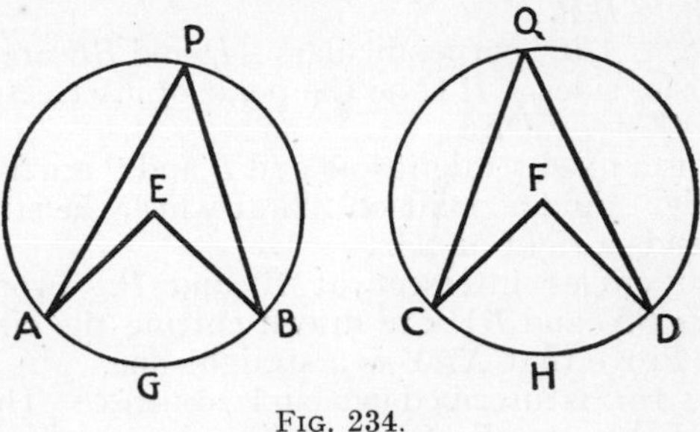

FIG. 234.

Given. Two equal circles *APBG, CQDH*, centres *E* and *F*, in which

$$\text{arc } AGB = \text{arc } CHD.$$

To prove. (1) $\angle AEB = \angle CFD$.
(2) $\angle APB = \angle CQP$.

Proof. (1) Apply the circle *APBG* to circle *CQDH*, so that *E* falls on *F*, and *EA* lies along *FC*.

Since radii of the circle are equal. $\therefore$ *A* falls on *C*.

Also, since arc *AGB* = arc *CHD*.

$\therefore$ circumferences coincide.

$\therefore$ *B* falls on *D*, and *EB* coincides with *FD*., *i.e.*, $\angle AEB$ coincides with $\angle CFD$.

$$\therefore \quad \angle AEB = \angle CFD.$$

(2) To prove $\angle$s *APB, CQD* equal.

Now $\angle APB = \frac{1}{2}\angle AEB$ (*Th.* 40)
and $\angle CQD = \frac{1}{2}\angle CFD$ (*Th.* 40)
But $\angle AEB = \angle CFD$ (*proved above*)
$\therefore \quad \angle APB = \angle CQD$.

Exercise 34

1. Prove that any parallelogram which is inscribed in a circle is a rectangle.

2. From a point O without a circle two straight lines OAB, OCD are drawn cutting the circumference in A, B, C, D. Prove $\angle OAD = \angle OCB$.

3. A $\triangle ABC$ is inscribed in a circle. O, the centre of the circle, is joined to D, the mid point of BC. Prove that $\angle BOD = \angle BAC$.

4. In a $\triangle ABC$, perpendiculars AD and BE are drawn to the opposite sides. If O be the point of intersection, prove that $\angle DOC = \angle DEC$.

5. AB is a fixed straight line and P and Q are fixed points without it. Find a point on AB at which the straight line PQ subtends a right angle.

6. Two circles intersect at A and B. From B two diameters BX and BY are drawn cutting the circles at X and Y. Prove that XAY is a straight line.

7. A $\triangle ABC$ is inscribed in a circle, centre O. The straight line which bisects the angle AOC, when produced, meets BC at D. Prove that $AODB$ is a cyclic quadrilateral.

8. In the $\triangle ABC$, AD is drawn perpendicular to BC. AE is a diameter of the circle ABC. Prove that the $\triangle$s ABD, AEC are equiangular.

SECTION 13

TANGENTS TO A CIRCLE

Definition. *A tangent to a circle is a straight line which meets the circle at one point, but being produced in either direction, does not meet it again* (see *Chap.* 19, *Part I*).

Theorem 48

(1) The straight line which is drawn perpendicular to a radius of a circle at the point where it meets the circumference is a tangent to the circle.

Conversely:

(2) A tangent to a circle is perpendicular to the radius drawn through the point of contact.

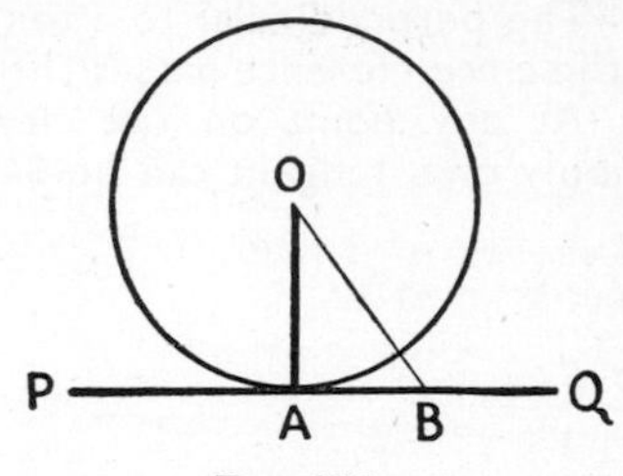

FIG. 235.

(1) *Given.* O is the centre of a circle and OA is a radius. The straight line PQ is perpendicular to OA at A.

To prove. PQ is a tangent to the circle.

Proof. Let B be any other point on PQ. Join OB.
In $\triangle OAB$, $\angle OAB$ is a right angle.
$\therefore$ $\angle OBA$ is less than a right angle.
$\therefore$ $OB > OA$ (*Th.* 18)
$\therefore$ B lies outside the circle.
Similarly it may be shown that any other point on PQ except A lies outside the circle.

∴ PQ meets the circumference at A and being produced does not cut it, *i.e.*, it does not meet it at another point.

∴ PQ is a tangent to the circle (*Def.*)

Converse:

(2) *Given.* PQ is a tangent to the circle at A.

To prove. PQ is perpendicular to OA.

Proof. If OA is not perpendicular to PQ draw OB perpendicular to it.

Then $\angle OBA$ is a right angle.

∴ $\angle OAB$ is less than a right angle.

∴ OB is less than OA.

But OA is a radius.

∴ B lies within the circle.

∴ AB if produced must cut the circle again.

But this is impossible since PQ is a tangent (*given*)

∴ OA must be perpendicular to PQ.

Corollary 1. The perpendicular to a tangent at its point of contact with the circumference passes through the centre.

Corollary 2. At any point on the circumference of a circle, one, and only one, tangent can be drawn.

Theorem 49

Common tangent

Definition. *For definitions see Part I*, § 141.

If two circles touch one another the straight line joining their centres, produced if necessary, passes through their point of contact.

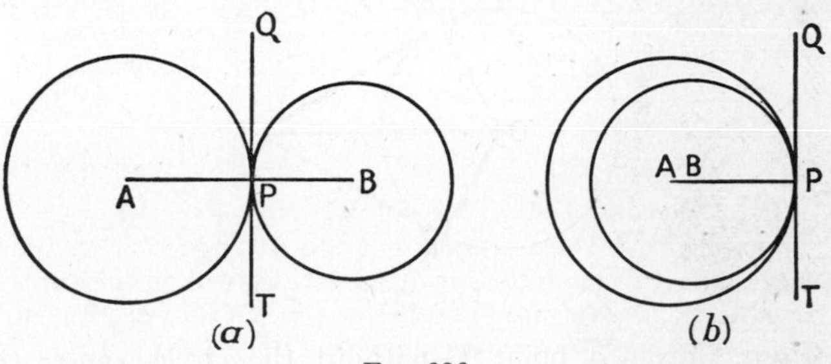

FIG. 236.

Given. Two circles, centres A and B touch at P.

The contact may be external as in Fig. 236 (*a*) or internal as in Fig. 236 (*b*).

To prove. A, P and B are in the same straight line.

Construction. Draw the common tangent QT.

Since the circles touch they have a common tangent.

Proof. AP and BP are each perpendicular to QT (*Th.* 48)

$\therefore$ $\angle$s QPA, QPB are right angles.

$\therefore$ AP and PB are in the same straight line (*Th.* 2)

i.e., **A, P and B are in the same straight line.**

Theorem 50

If two tangents are drawn to a circle from an external point.

(1) The tangents are equal.

(2) They subtend equal angles at the centre of the circle.

(3) They make equal angles with the straight line joining the external point to the centre.

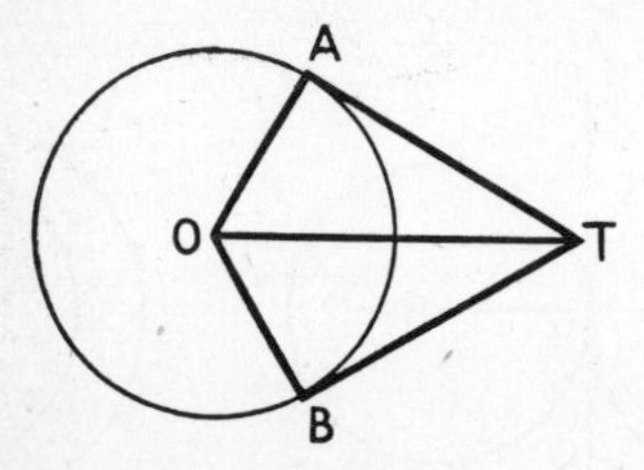

Fig. 237.

Given. From a point T without the circle, centre O, tangents TA, TB are drawn, touching the circle at A and B.

OA, OB, OT are joined.

To prove. (1) $TA = TB$

(2) $\angle AOT = \angle BOT$ (*angles subtended at* O)

(3) $\angle ATO = \angle BTO$.

Proof. In $\triangle$s TAO, TBO:

(1) $OB = OA$ (*radii*)

(2) OT is common

(3) $\angle OAT = \angle OBT$ (*right* $\angle$*s*, *Th.* 48)

$\therefore$ $\triangle$s TAO, TBO are congruent (*Th.* 17)

In particular:

(1) $TA = TB$.

(2) $\angle AOT = \angle BOT$.

(3) $\angle ATO = \angle BTO$.

Theorem 51

If a straight line touches a circle and, from the point of contact, a chord is drawn, the angles which the chord makes with the tangent are equal to the angles in the alternate segments of the circle.

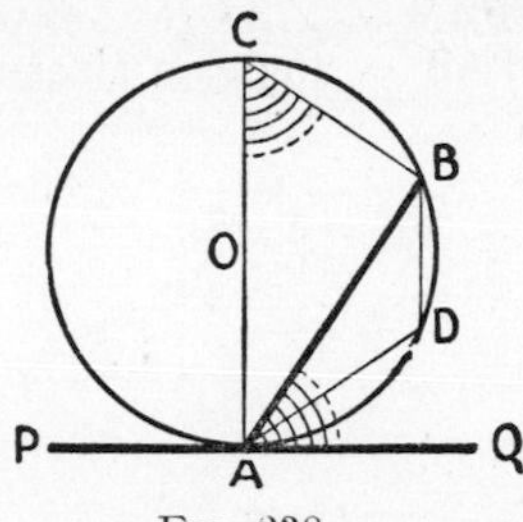

FIG. 238.

Given. PQ is a tangent to the circle ABC at A; AB is any chord.

To prove. (1) $\angle BAQ$ = angle in alternate segment ABC.
(2) $\angle BAP$ = angle in alternate segment ADB.

Construction. Draw the diameter AOC, O being the centre.
Take any point D on arc AB.
Join CB, BD, DA.

Proof. $\angle ACB$ = any other angle in segment ABC (*Th.* 41)
$\angle ADB$ = any other angle in segment ABD.

It is only necessary therefore to prove:

$$\angle BAQ = \angle ACB.$$
$$\angle BAP = \angle ADB.$$

(1) $\triangle ABC$ is right angled (*Th.* 43)

$\therefore$ $\angle BCA + \angle BAC$ = a right $\angle$ (*Th.* 10, *Cor.*)

but $\angle BAQ + \angle BAC$ = a right $\angle$. (*Th.* 48)

$\therefore$ $\angle BAQ = \angle BCA$.

$\therefore$ $\angle$BAQ = angle in alternate segment ABC.

(2) $ADBC$ is a cyclic quadrilateral.

$\therefore$ $\angle ADB + \angle ACB = 2$ right $\angle$s (*Th.* 44)

also $\angle PAB + \angle BAQ = 2$ right $\angle$s.

$\therefore$ $\angle PAB + \angle BAQ = \angle ADB + \angle ACB$.

But $\angle ACB = \angle BAQ$ (*proved above*)

Subtracting these equal angles

$\therefore$ $\angle PAB = \angle ADB$

i.e., $\angle$PAB = angle in alternate segment ADB.

Construction No. 17

To construct an exterior common tangent to two circles of unequal radii.

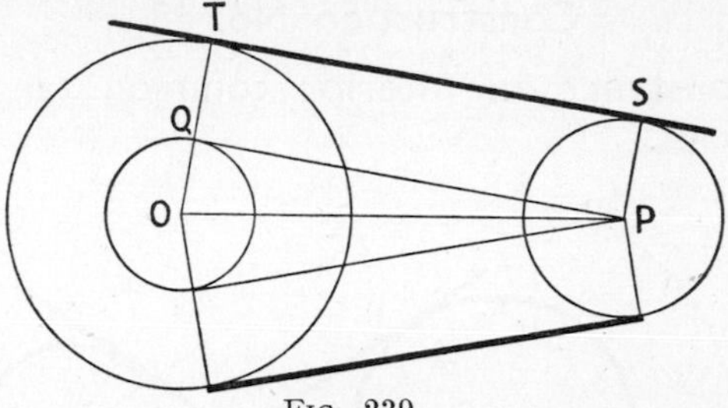

FIG. 239.

Given. Two circles, centres O and P, the circle with O as centre having the larger radius.

Required. To draw an exterior tangent to the circles.

Construction. Describe a circle, with centre O, and radius OQ equal to the **difference** of the two radii.
From P draw a tangent PQ to that circle.
Join OQ and produce it to meet the outer circle in T.
From P draw PS parallel to OT, meeting the circumference at S.
Join ST.

To prove. ST is a common tangent to the two circles.

Proof. Thus $OQ = OT - PS$

i.e., $OQ + PS = OT$

but $OQ + QT = OT$

$\therefore$ $PS = QT$

Also PS is parallel to QT.

$\therefore$ $PQTS$ is a parallelogram (*Th.* 25)

But $\angle PQT$ is a right $\angle$.

$\therefore$ $\angle$s QTS and PST are right $\angle$s.

I.e., ST is perpendicular to radii of the two circles at their extremities.

$\therefore$ **ST is a tangent to both circles.** (*Th.* 48)

Note.—Since two tangents can be drawn from P to the smaller circle, a second tangent to both circles may be constructed in the same way, on the other side as shown in Fig. 239.

Construction No. 18

To construct an interior common tangent to two circles.

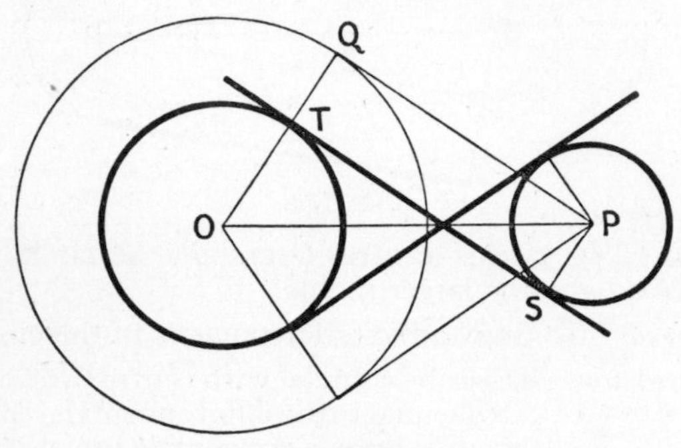

Fig. 240.

Given. Two circles, centres O and P.

Required. To draw an interior tangent to these two circles.

Construction. With centre O and radius equal to the sum of the two radii draw a circle.
From P draw a tangent, PQ, to this circle.
Join OQ, cutting the inner circle at T.
From P draw PS parallel to OQ.
Join ST.

ST is the required tangent.

The proof is similar to that of the exterior common tangent, Constr. No. 17, and two solutions are possible as shown in Fig. 240.

Construction No. 19

On a given straight line to construct a segment of a circle to contain a given angle.

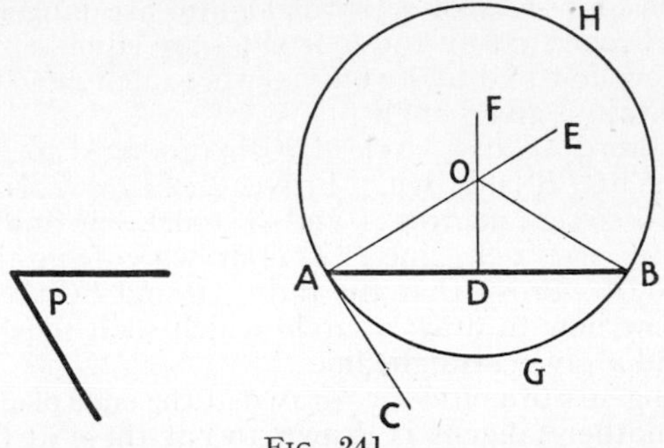

FIG. 241.

Given. AB is the given straight line and P the given angle.

Construction. At A construct the $\angle BAC$ equal to P.
From A draw AE perpendicular to AC.
Draw DF the perpendicular bisector of AB, cutting AE at O.
With centre O and radius OA describe the circle $AGBH$.
Then the segment AHB is the segment required.

Proof. Since O lies on the perpendicular bisector of AB

$$OB = OA. \quad (Th.\ 38)$$

$\therefore$ the circle passes through B, and AB is a chord of the circle.

Since $\angle OAC$ is a right angle, AC is a tangent to the circle.

$\therefore$ $\angle CAB =$ the angle in the alternate segment AHB (*Th.* 51)

but $\angle CAB = \angle P$. (*Constr.*)

$\therefore$ the segment AHB is the segment required.

Exercise 35

1. The four sides of a quadrilateral $ABCD$ are tangential to a circle. Prove that $AB + CD = BC + AD$.

2. The four sides of a parallelogram are tangential to a circle. Prove that all the four sides are equal.

3. Show how to draw two tangents to a circle so that they may contain a given angle.

4. A chord AC of a circle ABC is produced to P. From P a tangent PB is drawn. Prove $\angle PCB = \angle ABP$.

5. Two circles, centres A and B, touch one another at C. Through C a straight line PCQ is drawn cutting the circles at P and Q. Prove that the radii AP and BQ are parallel.

6. Show how to draw a circle which shall touch a given circle and a given straight line.

7. Tangents to a circle are drawn at the ends of a diameter AB. Another tangent is drawn to cut these at C and D. Prove that $CD = AC + BD$.

8. Two tangents, OA and OB, to a circle are at right angles to one another. AC is any chord of the circle and BD is drawn perpendicular to it. Prove that $BD = CD$.

9. From a point T without a circle, centre O, tangents TP and TQ are drawn touching the circle at P and Q. OP is produced to meet at R the straight line TR which is perpendicular to QT. Prove $OR = RT$.

10. Two circles touch internally at A. A chord BC of the larger circle is drawn to touch the inner circle at D. Prove that $\angle BAD = \angle CAD$.

SECTION 14

CONCURRENCIES CONNECTED WITH A TRIANGLE

1. Perpendicular bisectors of the sides

Theorem 52

The perpendicular bisectors of the three sides of a triangle are concurrent (see Part I, § 119).

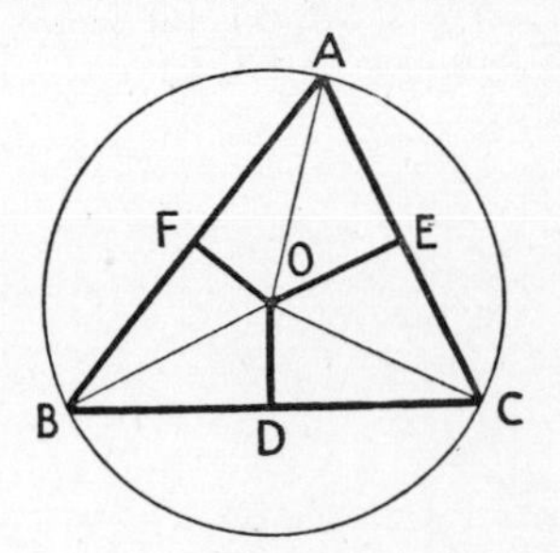

FIG. 242.

Given. OD and OE are the perpendicular bisectors of the sides BC and CA of the $\triangle ABC$, and they intersect at O.

Let F be the mid point of AB. Join OF.

To prove. OF is perpendicular to AB.

Proof. In $\triangle$s BDO, CDO:

(1) $BD = DC$ (*given*)
(2) OD is common.
(3) $\angle ODB = \angle ODC$ (*given*)

$\therefore$ $\triangle$s BDO, CDO are congruent.

In particular $OB = OC$.

Similarly from the $\triangle$s AEO, CEO, it may be proved

that $OC = OA$
but $OC = OB$ (*proved*)

$\therefore$ $OA = OB$.

In Δs AOF, BOF:

(1) $OA = OB$ (*proved*)
(2) OF is common.
(3) $AF = BF$ (*constr.*)

$\therefore$ Δs are congruent.

In particular $\angle OFA = \angle OFB$.

$\therefore$ OF is perpendicular to AD.

$\therefore$ the three perpendicular bisectors of the sides meet in O.

Note.—Since $OA = OB = OC$, O is the centre of the circumscribing circle of the ΔABC (see also Part I, § 49).

2. Bisectors of the angles

Theorem 53

The bisectors of the three angles of a triangle are concurrent.

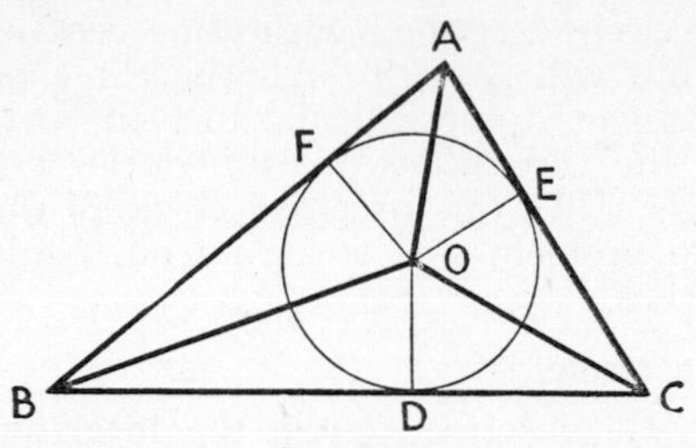

FIG. 243.

Given. OB and OC are the bisectors of the $\angle$s ABC, ACB of $\triangle ABC$.

Join OA.

To prove. OA bisects the angle BAC.

Construction. Draw OD, OE, OF perpendicular to BC, CA and AB, respectively.

Proof. In $\triangle$s ODC, OEC:

(1) $\angle OCD = \angle OCE$ (*given*)
(2) $\angle ODC = \angle OEC$ (*constr.*)
(3) OC is common.

$\therefore$ $\triangle$s ODC, OEC are congruent.

In particular $OD = OE$.

Similarly, $\triangle$s ODB, OFB may be proved congruent, and

$$OD = OF.$$

But $OD = OE$ (*proved*)

$\therefore$ OE = OF.

In $\triangle$s AOE, AOF:

(1) $OE = OF$ (*proved*)
(2) AO is common.
(3) $\angle AEO = \angle AFO$ (*right $\angle$s by constr.*)

$\therefore$ $\triangle$s are congruent.

In particular $\angle OAE = \angle OAF$.

$\therefore$ OA is the bisector of $\angle BAC$.

The inscribed and escribed circles of a triangle

In the above proof *OD, OE, OF* were proved equal. Therefore a circle described with *O* as centre, and one of them as radius will pass through the three points. Also, since each of them is perpendicular to a side at the extremity of the line, the three sides are tangential. The circle so constructed is called the **inscribed circle of the triangle.**

The whole problem was treated from a different point of view in § 142.

The escribed circles.

To obtain the inscribed circle the **interior angles** were

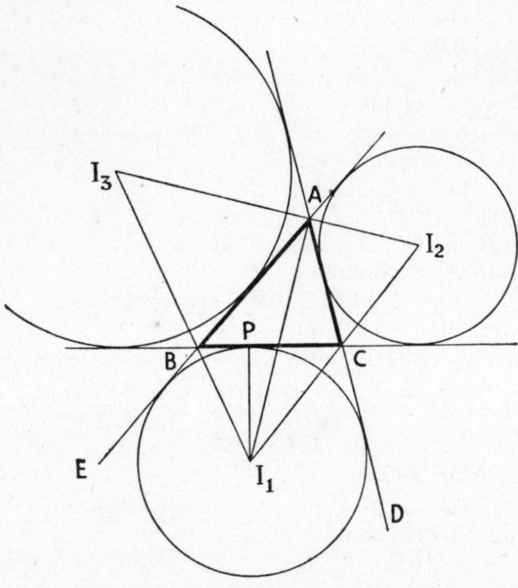

Fig. 244.

bisected and the bisectors were concurrent. Let us now construct the **bisectors of the exterior angles.** Produce the sides AB and AC through suitable distances to D and E.

Bisect the exterior $\angle$s EBC, BCD. Let the bisectors meet at I_1. Draw I_1P perpendicular to BC. The length of this can be proved, as above, to be equal to perpendiculars drawn from I_1 to BE and CD.

If, therefore, a circle be described with I_1 as centre, and I_1P as radius it will touch BC, and the other two sides produced. This circle is an **escribed circle.** Similar circles can be described by bisecting other exterior angles and I_2, I_3 are the centres of two other escribed circles. There are thus **three escribed circles** to a triangle. It may easily be proved that AI_1 bisects the interior angle BAC. Consequently each of the centres I_1, I_2, I_3 is the intersection of the bisectors of two exterior angles and the opposite interior angle.

The student, as an exercise, should prove that I_1I_2, I_2I_3, I_3I_1, are straight lines and form a triangle.

3. Medians

Theorem 54

The three medians of a triangle are concurrent.

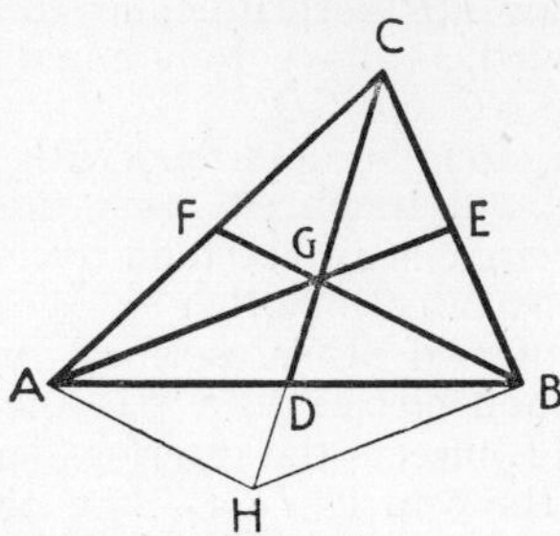

FIG. 245.

Given. E and F are the mid points of the sides BC, AC of the $\triangle ABC$.

The straight lines AE, BF intersect at G.

Join CG and produce it to meet AB at D.

To prove. D is the mid point of AB, *i.e.*, $AD = DB$.

Construction. Produce CD to H making $GH = CG$.
Join AH, BH.

Proof. In the $\triangle ACH$, FG joins the mid points of the sides AC, HC.

$\therefore$ FG is parallel to AH (*Th.* 27)

i.e., GB is parallel to AH.

Similarly from the $\triangle BCH$, it may be shown that

AG is parallel to BH.

$\therefore$ **AGBH is a parallelogram.**

$\therefore$ **AD = DB** (*Th.* 23)

Notes. (1) The point G is called the **centroid** of $\triangle ABC$. It is the centre of gravity of a triangular lamina in the form of the $\triangle ABC$.

(2) Since $GD = \frac{1}{2}GH$.
Then $GD = \frac{1}{2}CG$.
$\therefore$ **GD = $\frac{1}{3}$CD.**
Similarly $EG = \frac{1}{3}AE$ and $FG = \frac{1}{3}BF$.

4. Altitudes

Theorem 55

The perpendiculars drawn from the vertices of a triangle to the opposite sides are concurrent.

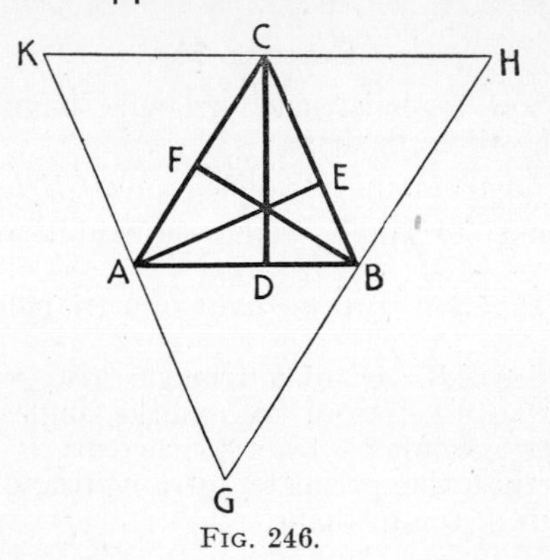

Fig. 246.

Given. In $\triangle ABC$, AE, BF, CD are perpendiculars drawn from the vertices A, B, C to the opposite sides, *i.e.*, they are altitudes drawn from the three vertices.

To prove. AE, BF, CD are concurrent.

Construction. Through the vertices A, B, C draw straight lines parallel to the opposite sides CB, AC, AB respectively, to form the $\triangle GHK$.

Proof. Quadrilateral $GACB$ is a parallelogram (*Constr.*)

$\therefore \quad AG = CB$ (*Th.* 22)

Similarly $ABCK$ is a parallelogram

and $KA = CB$

$\therefore \quad KA = AG.$

Since CB and KG are parallel and AE is perpendicular to CB.

$\therefore$ AE is the perpendicular bisector of GK.

Similarly BF and CD are the perpendicular bisectors of GH and KH respectively.

I.e., AE, BF and CD are the perpendicular bisectors of the sides of the $\triangle GHK$.

$\therefore$ **AE, BF and CD are concurrent** (*Th.* 52)

Exercise 36

1. Show how to construct a triangle, having given the lengths of the three medians.

[*Hint.*—Construct the parallelogram $AGBH$ in Fig. 245.]

2. Construct a triangle, given the middle points of the sides.

3. Prove that any two medians of a triangle are greater than the third.

4. The sides AB, AC of a triangle ABC are produced. Show that the bisectors of the exterior angles at B and C and the interior angle at A are concurrent.

5. Prove that the perimeter of any triangle is greater than the sum of the medians.

SECTION 15

RATIO IN GEOMETRY

Division of a straight line in a given ratio

In Part I, Chap. 20, the meaning of a ratio in connection with the comparison of the lengths of straight lines was briefly examined: it was pointed out that by the ratio of the lengths of two lines is meant the ratio of the numbers which express the measures of the lengths of the lines **in terms of the same units.** If therefore a straight line AB

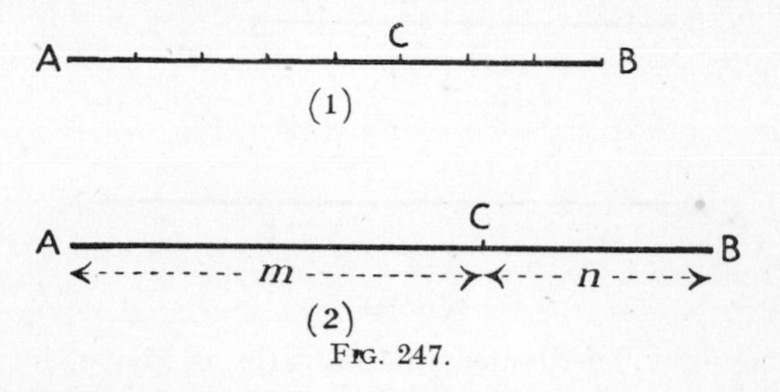

FIG. 247.

(Fig. 247 (1)) be divided into two parts at C and AC contains 5 units of length while BC contains 3 of the same units, the ratio of the lengths of the two lines is 5 : 3. The whole line AB contains $5 + 3$, *i.e.*, 8 of these units, all of them equal to one another. Thus if the whole line be divided into 8 equal parts AC will contain 5 of them while CB contains 3. This may be expressed in the form

$$\frac{AC}{BC} = \frac{5}{3}.$$

In general, if in Fig. 247 (2) AC contains m units of length and BC n units, then the ratio $\frac{AC}{BC} = \frac{m}{n}$. Again if AB be divided into $m + n$ equal parts, AC will contain m of these and BC n.

The division of a line in a given ratio may be extended. Thus, in Fig. 248 (1) the point P is said to **divide AB internally** in the ratio $AP : PB$.

If the line be produced to any point P (Fig. 250 (2)) the line AB is said **to be divided externally in the ratio AP : PB.**

It will be noticed that in each case **the segments of the line are measured from P, the point of division, to the two ends of the line, A and B.**

This division of a line into m or n equal parts is readily comprehended when m and n are integers. It is equally so if m and n are exact decimal fractions. For example if $m = 3{\cdot}4$ and $n = 1{\cdot}3$, so that $m + n = 4{\cdot}7$, we may express the ratio $3{\cdot}4 : 1{\cdot}3$ in the equivalent form of $34 : 13$. Thus

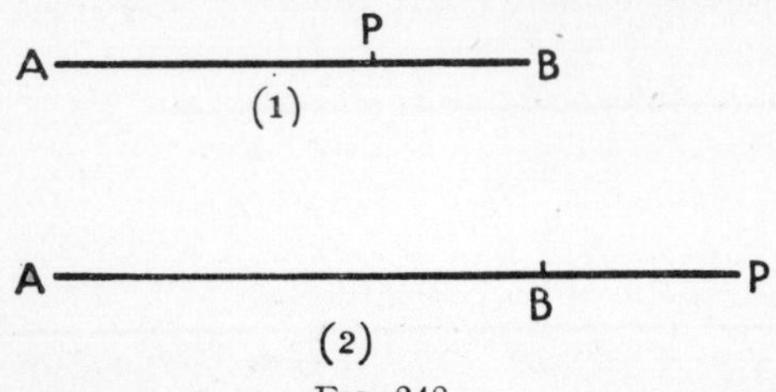

FIG. 248.

the lines may be divided in this ratio as above, but the units will be each one-tenth of those previously employed. The reasoning remains the same.

Incommensurables. But cases arise when it is not possible to express the ratio of two quantities by the use of exact, definite integers.

For example if the side of a square be a units of length, we know, by the use of the Theorem of Pythagoras that a diagonal is $a\sqrt{2}$ units of length (see § 102, Part I). Thus

the ratio of a diagonal of a square to a side is $\sqrt{2} : 1$.

Now, $\sqrt{2}$ is what is called an **irrational number,** *i.e.*, it cannot be expressed exactly in the form of a fraction or a decimal. Its value can be found by arithmetic to, say, four significant figures when it is $1{\cdot}414$, but there is no limit to the number of places to which it may be worked. Thus the ratio $\sqrt{2} : 1$ can be expressed as $1{\cdot}414 : 1$ approximately. Consequently a **straight line cannot be divided exactly into $\sqrt{2}$ equal parts.**

Similarly the ratio of the circumference of a circle to its diameter is denoted by the symbol π (see § 122, Part I). But the value of π **cannot be determined exactly** though it may be found to any required degree of accuracy. To 6 significant figures $\pi = 3{\cdot}14159\ldots$

Quantities such as in the above examples, the ratios of which cannot be expressed by exact integers, are said to be **incommensurable.**

In considering proofs involving ratios, it has hitherto been assumed that the numbers used are not incommensurables, and the same assumption will be made in proofs which follow in this section. Otherwise these would be too difficult for the student at this stage in his study of mathematics.

Geometrical representation of irrational numbers

It is of interest to note that, in the case of an irrational number such as $\sqrt{2}$, although it cannot be expressed arithmetically in an exact form, it is possible theoretically to represent it **exactly** by the length of a straight line. In

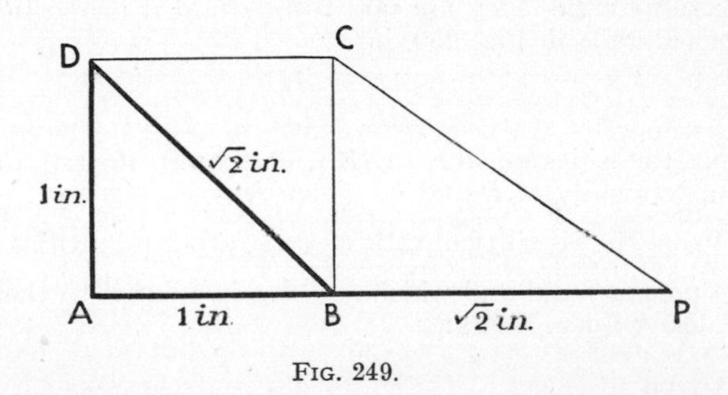

Fig. 249.

Fig. 249, $ABCD$ is a square whose side is one inch in length. Then the length of the diagonal BD is $\sqrt{2}$ in. If AB be produced to P and BP be made equal to BD, then the length of BP is $\sqrt{2}$ inches. Thus

$$BP : BA = \sqrt{2} : 1$$

or AP is divided at B in the ratio $\sqrt{2} : 1$. The exactitude secured depends on the accuracy of the instruments used and the skill of the draughtsman.

Straight lines which represent other irrational numbers may similarly be obtained by the use of right-angled triangles and the application of the Theorem of Pythagoras to them. For example, in Fig. 249 the length of CP is $\sqrt{3}$ inches.

But, **it is not possible to obtain a straight line the length of which is exactly π units.**

Proportional. If each of two straight lines is divided in the same ratio, they are said to be divided proportionally.

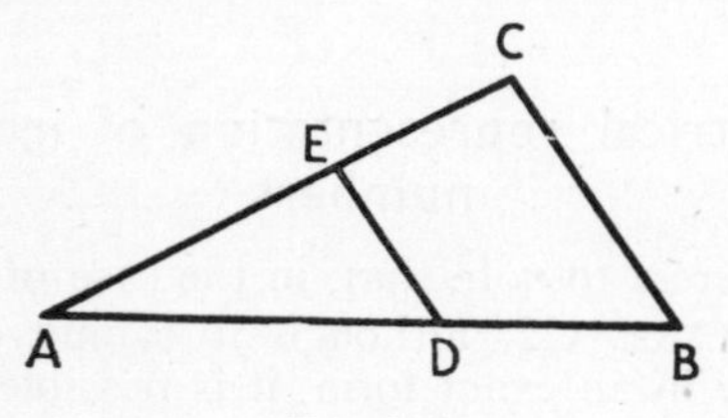

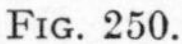
FIG. 250.

For example in Fig. 250, if

$$\frac{AD}{DB} = \frac{AE}{EC}$$

then the sides of the $\triangle ABC$, AB and AC are divided proportionally at D and E.

From the equality of ratios $\frac{a}{b} = \frac{c}{d}$, other quantities which are proportional and which are closely related to them can be derived, as follows:

Given $\qquad \frac{a}{b} = \frac{c}{d}.$

Then $\qquad (1)\ \frac{b}{a} = \frac{d}{c},$

also (2) $\frac{a}{b} + 1 = \frac{c}{d} + 1$

or $$\frac{a+b}{b} = \frac{c+d}{d}.$$

Similarly, (3) $\frac{a-b}{b} = \frac{c-d}{d}$

and (4) $\frac{a+b}{a-b} = \frac{c+d}{c-d}.$

These may be illustrated from Fig. 250.

Given $$\frac{AD}{DB} = \frac{AE}{EC}.$$

Then by (1) $$\frac{DB}{AD} = \frac{EC}{AE}$$

and by (2) $$\frac{AD + DB}{DB} = \frac{AE + EC}{EC}$$

or $$\frac{AB}{DB} = \frac{AC}{EC}.$$

Similarly, $$\frac{AB}{AD} = \frac{AC}{AE}.$$

Mean proportional. If a, b, c are numbers such that

$$\frac{a}{b} = \frac{b}{c}$$

or $$ac = b^2$$

or $$b = \sqrt{ae}.$$

Then **b** is called a **mean proportional** between **a** and **c**.

Theorem 56

If a straight line is drawn parallel to one side of a triangle, the other sides are divided proportionally.

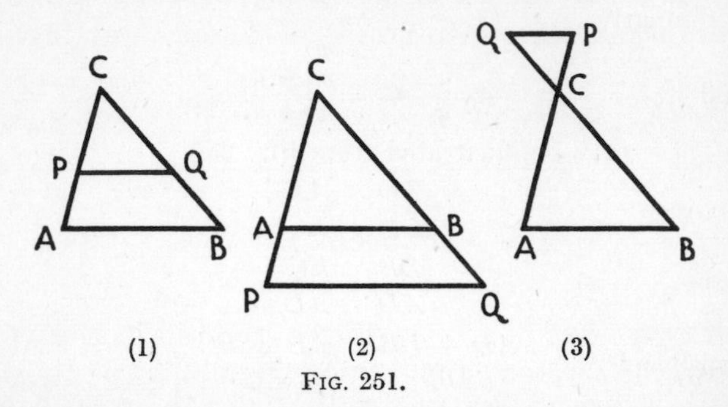

FIG. 251.

Given. ABC is a triangle in which PQ is drawn parallel to AB cutting AC in P and BC in Q, either internally as in (1) or externally as in (2) or (3).

To prove.
$$\frac{PC}{PA} = \frac{QC}{QB}.$$

Proof. Let P divide CA in the ratio $m : n$,

i.e.,
$$\frac{PC}{PA} = \frac{m}{n}.$$

Let PC be divided into m equal parts.
Let PA be divided into n equal parts.
Then the parts in PC and PA are all equal. (See p. 303.)
From each point thus obtained on PC and PA, suppose a straight line be drawn parallel to AB.
These lines thus divide QC into m equal parts and QB into n equal parts.
Each of the parts in QC is equal to each of the parts in QB. (*Th.* 28.)

$$\therefore \quad \frac{QC}{QB} = \frac{m}{n}$$

but
$$\frac{PC}{PA} = \frac{m}{n}.$$
$$\therefore \quad \frac{PC}{PA} = \frac{QC}{QB}.$$

Thus CA and CB are divided proportionally at P and Q.

Corollary. It may be proved in the same way that

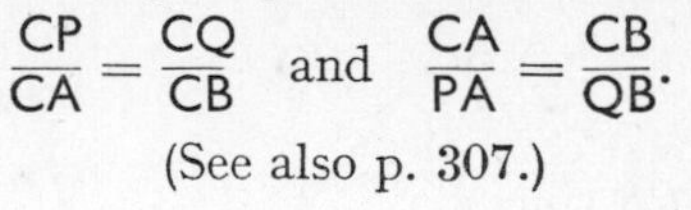

$$\frac{CP}{CA} = \frac{CQ}{CB} \quad \text{and} \quad \frac{CA}{PA} = \frac{CB}{QB}.$$

(See also p. 307.)

Theorem 57

(Converse of Theorem 56)

If two sides of a triangle are divided in the same ratio, the straight line joining the points of division is parallel to the third side.

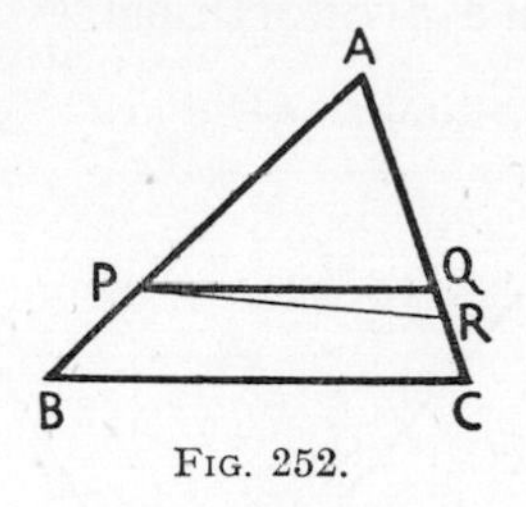

FIG. 252.

Given. In $\triangle ABC$, $\frac{PA}{PB} = \frac{QA}{QC}$.

To prove. PQ is parallel to BC.

Construction. Draw PR parallel to BC.

Proof.
$$\frac{PA}{PB} = \frac{QA}{QC} \qquad \textit{(given)}$$
$$\therefore \quad \frac{AB}{PB} = \frac{AC}{QC} \quad (1) \qquad \textit{(Th. 56, Cor.)}$$

Since PR is parallel to BC.

$$\therefore \quad \frac{AB}{PB} = \frac{AC}{RC} \quad (2) \qquad (Th.\ 56)$$

Comparing (1) and (2).

$$\frac{AC}{QC} = \frac{AC}{RC}.$$

$$\therefore \quad QC = RC.$$

$\therefore$ R and Q must coincide.
Thus PQ and PR coincide.

But PR is parallel to BC (*constr.*)

$\therefore$ **PQ is parallel to BC.**

Theorem 58

If two triangles are equiangular, their corresponding sides are proportional.

Such triangles are similar (see Part I, §§ 145–147).

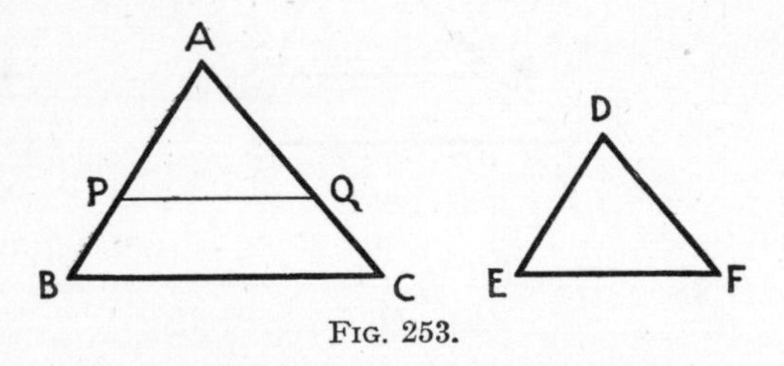

FIG. 253.

Given. ABC, DEF are equiangular having $\angle A = \angle D$, $\angle B = \angle E$, $\angle C = \angle F$.

To prove. The sides of the Δs are proportional, *i.e.*,

$$\frac{AB}{DE} = \frac{BC}{EF} = \frac{CA}{FD}.$$

Construction. From AB cut off $AP = DE$, and from AC, $AQ = DF$.

Join PQ.

Proof. In $\triangle$s APQ, DEF:

(1) $AP = DE$ (*constr.*)
(2) $AQ = DF$ (*constr.*)
(3) $\angle A = \angle D$ (*given*)

$\therefore$ $\triangle$s APQ, DEF are congruent.

$$\therefore \quad \angle APQ = \angle DEF.$$

But $\angle ABC = \angle DEF$ (*given*)

$$\therefore \quad \angle APQ = \angle ABC.$$

$\therefore$ PQ is parallel to BC (*Th.* 7)

$$\therefore \quad \frac{AB}{AP} = \frac{AC}{AQ} \qquad (Th.\ 56,\ Cor.)$$

But $AP = DE$ and $AQ = DF$ (*proved*)

$$\therefore \quad \frac{AB}{DE} = \frac{AC}{DF}.$$

Similarly it may be proved that

$$\frac{AC}{DF} = \frac{BC}{EF}.$$

$\therefore$ the ratios $\frac{AB}{DE}, \frac{BC}{EF}, \frac{CA}{FD}$ are all equal.

$\therefore$ $\triangle$s ABC, DEF are similar.

Theorem 59

(*Converse of Theorem* 58)

If the corresponding sides of two triangles are proportional the triangles are equiangular and therefore similar.

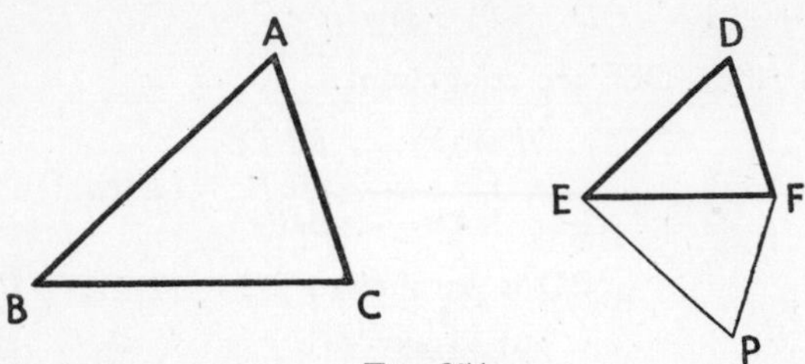

FIG. 254.

Given. In $\triangle$s ABC, DEF, $\frac{AB}{DE} = \frac{BC}{EF} = \frac{AC}{DF}$.

To prove. $\angle A = \angle D$, $\angle B = \angle E$, $\angle C = \angle F$.

Construction. At E and F, on the side opposite from D make $\angle FEP = \angle ABC$, $\angle EFP = \angle ACB$; then EP and FP meeting at P form the $\triangle PEF$.

Proof. $\triangle$s ABC, PEF are equiangular by construction.

$$\therefore \quad \frac{AB}{PE} = \frac{BC}{EF} \quad (Th.\ 58)$$

but $$\frac{AB}{DE} = \frac{BC}{EF} \quad (given)$$

$$\therefore \quad \frac{AB}{PE} = \frac{AB}{DE}.$$

$$\therefore \quad PE = DE.$$

Similarly, $$PF = DF.$$

In $\triangle$s DEF, PEF:

(1) $PE = DE$.
(2) $PF = DF$.
(3) EF is common.

$\therefore$ $\triangle$s DEF, PEF are congruent. (*Th.* 13)

In particular $\angle DEF = \angle PEF$
$\angle DFE = \angle PFE$.
But $\angle PEF = \angle ABC$ (*constr.*)
$\therefore$ $\angle DEF = \angle ABC$.
Similarly, $\angle PFE = \angle ACB$
$\therefore$ $\angle DFE = \angle ACB$
and remaining $\angle EDF = \angle BAC$.
$\therefore$ $\triangle$s ABC, DEF are equiangular.

Theorem 60

If two triangles have an angle of one triangle equal to an angle of the other and the sides about these equal angles proportional, the triangles are equiangular.

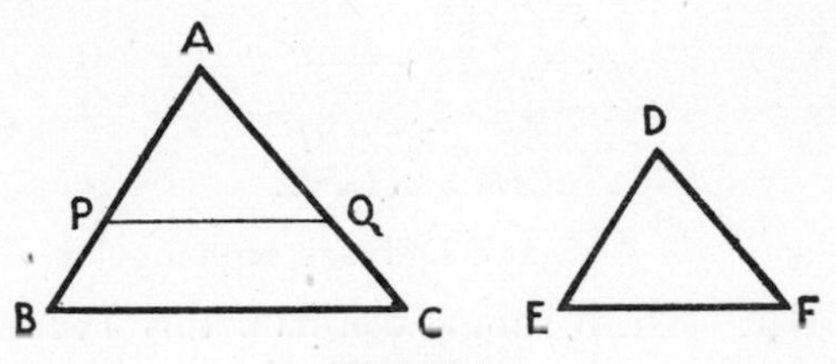

FIG. 255.

Given. In $\triangle$s ABC, DEF,

$$\angle BAC = \angle EDF \quad \text{and} \quad \frac{AB}{DE} = \frac{AC}{DF}.$$

To prove. $\triangle$s ABC, DEF are equiangular,

i.e., $\angle B = \angle E$, $\angle C = \angle F$.

Construction. From AB, AC, cut off AP and AQ equal respectively to DE and DF.
Join PQ.

Proof. In $\triangle$s APQ, DEF:

(1) $AP = DE$ (*constr.*)
(2) $AQ = DF$ (*constr.*)
(3) $\angle PAQ = \angle EDF$ (*given*)

$\therefore$ $\triangle$s APQ, DEF are congruent. (*Th.* 4)

In particular $\angle APQ = \angle DEF$
$\angle AQP = \angle DFE.$

Since $\dfrac{AB}{DE} = \dfrac{AC}{DF}$ (*given*)

and $DE = AP$ (*constr.*)
also $DF = AQ.$ (*constr.*)

$$\therefore \frac{AB}{AP} = \frac{AC}{AQ}.$$

$\therefore$ PQ is parallel to BC (*Th.* 57)

$\therefore$ $\angle APQ = \angle ABC$
and $\angle AQP = \angle ACB.$
But $\angle APQ = \angle DEF$ (*proved*)
and $\angle AQP = \angle DFE$ (*proved*)
$\therefore$ $\angle ABC = \angle DEF$
and $\angle ACB = \angle DFE.$

$\therefore$ $\triangle$s ABC and DEF are equiangular.

Note. The student should compare this Theorem with Theorem 4.

Theorem 61

In a right-angled triangle, if a perpendicular is drawn from the right angle to the hypotenuse, the triangles on each side of the perpendicular are similar to the whole triangle and to each other.

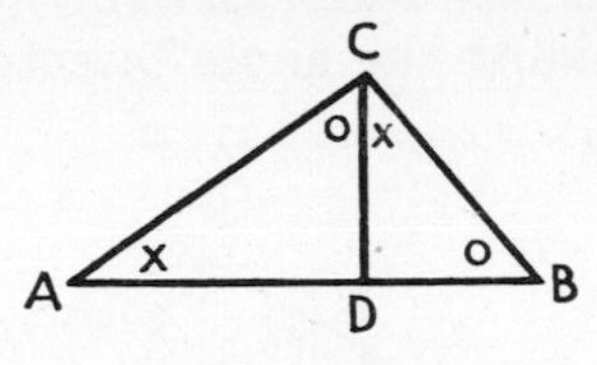

FIG. 256.

Given. $\triangle ABC$, right angled at C.
CD is perpendicular to the hypotenuse AB.

To prove. $\triangle$s CAD, CDB, ABC are similar.

Proof. In $\triangle$s ACB, ADC,

$$\angle ACB = \angle ADC \quad (right\ \angle s)$$

$$\angle CAD \text{ is common.}$$

$$\therefore \quad \angle ABC = \angle ACD. \quad (Th.\ 10)$$

$\therefore$ $\triangle$s ACB, ADC are equiangular and similar.

Similarly it may be proved that

$\triangle$s ACB, DCB are equiangular and similar.

$\therefore$ the three $\triangle$s CAD, CDB, ABC are similar.

Corollary 1. The square on the perpendicular is equal to the rectangle contained by the segments of the base.

Since $\triangle$s ACD, BCD (Fig. 256) are similar.

$$\therefore \quad \frac{AD}{CD} = \frac{CD}{DB}.$$

$$\therefore \quad AD \,.\, DB = CD^2.$$

Corollary 2. The perpendicular CD is a mean proportional between the segments of the base.

Theorem 62

(*a*) The *internal* bisector of an angle of a triangle divides the opposite side *internally* in the ratio of the sides containing the angle bisected.

(*b*) The *external* bisector of an angle of a triangle divides the opposite side *externally* in the ratio of the sides containing the angle bisected.

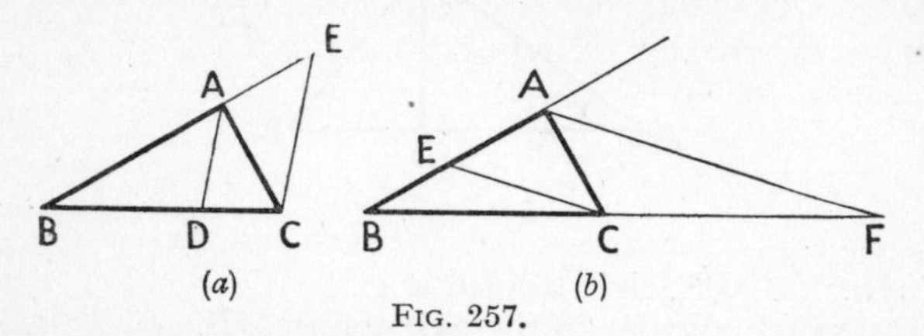

Fig. 257.

(a) *Given.* AD bisects an *internal* angle BAC of the $\triangle ABC$, and meets BC at D (Fig. 257 (*a*)).

To prove. $$\frac{BD}{DC} = \frac{BA}{AC}.$$

Construction. From C draw CE parallel to AD to meet BA produced in E.

Proof. EC is parallel to AD, and BE cuts them.

$$\therefore \quad \angle BAD = \angle BEC \quad (\textit{corr. } \angle s)$$

Also EC is parallel to AD and AC cuts them.

$$\therefore \quad \angle DAC = \angle ACE \quad (\textit{alt. } \angle s)$$

But $$\angle BAD = \angle DAC \quad (\textit{given})$$

$$\therefore \quad \angle BEC = \angle ACE.$$

$$\therefore \quad AE = AC.$$

Since AD is parallel to EC.

$$\therefore \quad \frac{BA}{AE} = \frac{BD}{DC} \quad (\textit{Th. } 56)$$

and $$AE = AC.$$

$$\therefore \quad \frac{\mathrm{BA}}{\mathrm{AC}} = \frac{\mathrm{BD}}{\mathrm{DC}}.$$

(b) *Given.* In Fig. 257 (*b*), AF bisects the external $\angle CAP$ and meets BC produced at F.

To prove. $$\frac{FB}{FC} = \frac{BA}{AC}.$$

Proof. The proof follows the same method as in (*a*) by drawing EC parallel to AF and proving $AE = AC$.

Then in $\triangle BAF$,

$$\frac{FB}{FC} = \frac{AB}{AE},$$

and since $AE = AC$,

$$\therefore \quad \frac{\mathrm{FB}}{\mathrm{FC}} = \frac{\mathrm{AB}}{\mathrm{AC}}.$$

Theorem 63

If two chords of a circle intersect, within or without the circle, the rectangle contained by the segments of one is equal to the rectangle contained by the segments of the other.

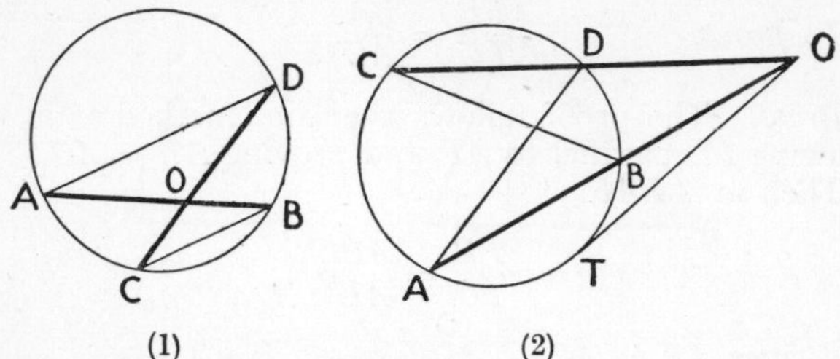

FIG. 258.

Given. The chords AB, CD, intersect at O within the circle in Fig. 258 (1) and without in Fig. 258 (2). The latter is a case of external division (see p. 304).

To prove. Rect. $OA \cdot OB =$ rect. $OC \cdot OD$ (in both cases)

Construction. Join BC, AD.

Proof. In $\triangle$s AOD, COB:

$\angle OAD = \angle OCB$ (*Th.* 41)
$\angle AOD = \angle COB$ in (1), same angle in (2).

$\therefore$ third angles in each are equal, *i.e.*, $\triangle$s AOD, COB are equiangular and similar.

$$\therefore \frac{OA}{OC} = \frac{OD}{OB}.$$

$$\therefore \text{rect. } OA \cdot OB = \text{rect. } OC \cdot OD.$$

Corollary. In (2) it may be noticed that if OBA were to rotate in an anti-clockwise direction about O, A and B would approach one another. Ultimately OBA would become the tangent OT.

Then the rect. $OA \cdot OB$ becomes OT^2,

i.e., rect. $OC \cdot OD = OT^2$.

Theorem 64

The areas of similar triangles are proportional to the squares of corresponding sides.

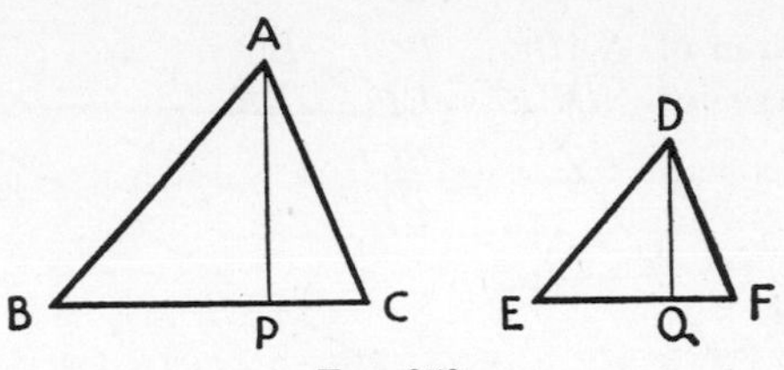

Fig. 259.

Given. $\triangle$s ABC, DEF are similar.

To prove. $\dfrac{\text{Area of } \triangle ABC}{\text{Area of } \triangle DEF} = \dfrac{BC^2}{EF^2}.$

Construction. Draw AP perpendicular to BC and DQ perpendicular to EF.
These are the altitudes corresponding to the sides BC, EF.

Proof. In $\triangle$s ABP, DEQ :

(1) $\angle ABP = \angle DEQ$ (*given*)
(2) $\angle APB = \angle DQE$ (*right* $\angle$*s*)
$\therefore$ (3) $\angle BAP = \angle EDQ$ (*Th.* 10)

$\therefore$ $\triangle$s ABP, DEQ are equiangular and similar.

$$\therefore \frac{AP}{DQ} = \frac{AB}{DE},$$

but $$\frac{AB}{DE} = \frac{BC}{EF} \quad (\textit{given, similar } \triangle s)$$

$$\therefore \frac{AP}{DQ} = \frac{BC}{EF}.$$

Now, $\dfrac{\text{Area of } \triangle ABC}{\text{Area of } \triangle DEF} = \dfrac{\frac{1}{2}BC \, . \, AP}{\frac{1}{2}EF \, . \, DQ}$ (*Th.* 30, *Cor.* 1)

$$= \frac{BC}{EF} \times \frac{AP}{DQ}.$$

But $\dfrac{AP}{DQ} = \dfrac{BC}{EF}$ (*proved*)

$$\therefore \quad \frac{\text{Area of } \triangle ABC}{\text{Area of } \triangle DEF} = \frac{BC}{EF} \times \frac{BC}{EF}$$

$$= \frac{BC^2}{EF^2}.$$

Exercise 37

1. $ABCD$ is a quadrilateral in which AB is parallel to CD. The diagonals intersect at P. Prove

$$\frac{AP}{AC} = \frac{BP}{BD}.$$

2. Three straight lines OP, OQ, OR are cut by two parallel straight lines in A, B, C and D, E, F, respectively. Prove that

$$\frac{AB}{BC} = \frac{DE}{EF}.$$

3. $ABCD$ is a parallelogram and DC is bisected at E. If BE meets AC in O, prove that $OC = \frac{1}{2}AO$.

4. ABC is a right-angled triangle with the right angle at A. From A draw AD, perpendicular to BC. Prove

$$\frac{BD}{DC} = \frac{AB^2}{AC^2}.$$

5. In the $\triangle ABC$ a straight line is drawn from A to meet BC in D. On AD any point O is taken. Prove

$$\frac{\triangle AOB}{\triangle AOC} = \frac{BD}{DC}.$$

6. ABC is a triangle and BE, CF are altitudes drawn from B and C respectively. Prove that

$$\frac{FE}{BC} = \frac{AF}{AC}.$$

7. C is a point on a straight line AB. On AC and CB equilateral $\triangle$s ACD, CBE are constructed on the same side of AB. The straight line joining D and E is produced to meet AB produced at F. Prove

$$\frac{FB}{BC} = \frac{FC}{CA}.$$

8. AB is a chord of a circle; AT and BT are tangents drawn from A and B. If O is the centre of the circle prove that

$$\frac{\text{Area of } \triangle ABT}{\text{Area of } \triangle OAB} = \frac{AT^2}{OA^2}.$$

SECTION 16

CONSTRUCTIONS

Construction No. 20

To construct a fourth proportional to three given straight lines.

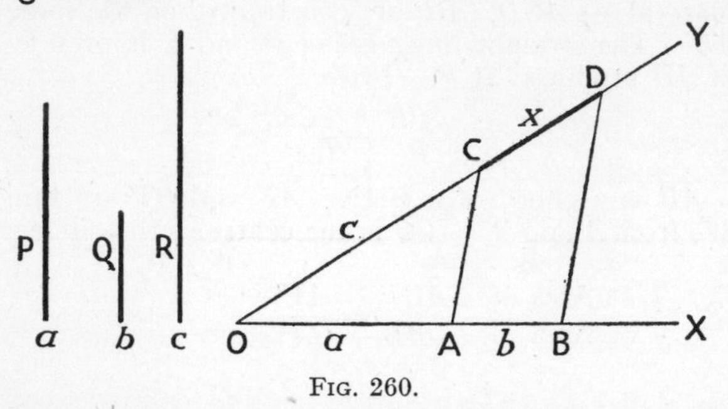

FIG. 260.

Given. a, b, c are the measures of the lengths of three straight lines P, Q, R.

Required. To find a line whose length, x, is such that

$$\frac{a}{b} = \frac{c}{x}.$$

Construction. Draw two straight lines OX, OY, of suitable length and containing any suitable angle.

Along OX mark $OA = a$ units and $AB = b$ units.
Along OY mark $OC = c$ units.
Join AC.
From B draw BD parallel to AC.
Then CD will represent x units.

Proof. Since AC and BD are parallel.

$$\therefore \quad \frac{OA}{AB} = \frac{OC}{CD} \qquad (Th. 56)$$

i.e.,

$$\frac{a}{b} = \frac{c}{x}.$$

$\therefore$ the length of CD is the fourth proportional required.

Construction No. 21

To divide a given straight line internally and externally in a given ratio.

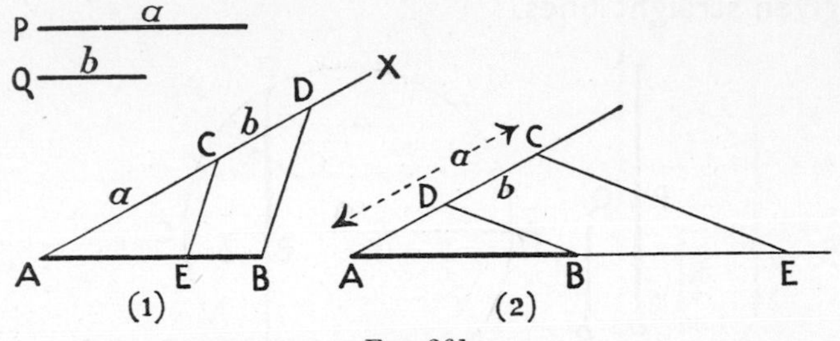

FIG. 261.

Given. AB is a straight line which it is required to divide (1) internally and (2) externally in the ratio of the lines P and Q of lengths a and b.

Construction. From A draw a straight line AX of indefinite length making any angle with AB. From AX cut off AC equal to P, *i.e.*, a units in length.

(1) For **internal** division from CX mark off $CD = b$ units.

(2) For **external** division from CA mark off $CD = b$ units.

In each case join DB.

From C draw CE parallel to DB to meet AB at E, internally in Fig. 261 (1) and externally in Fig. 261 (2).

E is the point of division in each case.

Proof. In each figure EC is parallel to the side BD of the $\triangle ABD$.

$$\therefore \quad AE : EB = AC : CD \quad (Th.\ 56)$$

But $$AC : CD = a : b.$$

$$\therefore \quad \mathbf{AE : EB = a : b.}$$

Construction No. 22

To construct a straight line whose length is the mean proportional between the lengths of two given straight lines.

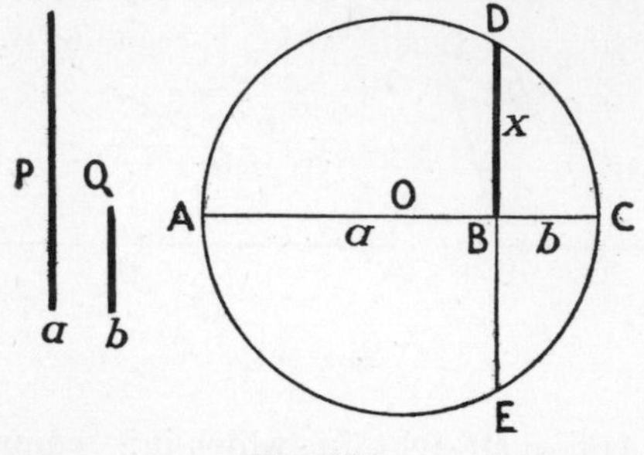

FIG. 262.

Given. P and Q are straight lines whose lengths are a and b units.

Required. To find a straight line of length x units such that

$$\frac{a}{x} = \frac{x}{b} \quad \text{or} \quad x^2 = ab.$$

Construction. Draw a straight line of suitable length.
Along it mark off $AB = a$ units, $BC = b$ units.
On AC as diameter draw a circle, centre O.
Through B draw a chord DE perpendicular to AC.

BD is the line required.

Proof. Rect. $AB \,.\, BC =$ rect. $BD \,.\, BE$ (*Th.* 63)

i.e., $$\frac{AB}{BD} = \frac{BE}{BC}.$$

But $$BD = BE.$$

$$\therefore \quad \frac{AB}{BD} = \frac{BD}{BC},$$

and $$AB \, . \, BC = BD^2$$

or $$\frac{a}{x} = \frac{x}{b} \quad \text{and} \quad x^2 = ab.$$

Corollary. This construction also provides the solution of the problem:

To construct a square equal to a given rectangle.

BD is the side of a square which is equal to the rectangle whose adjacent sides are AB and BC.

APPENDIX A

GEOMETRICAL REPRESENTATION OF ALGEBRAICAL IDENTITIES

The following algebraical identities, which in preceding pages have been employed in the proofs of geometrical theorems, can be illustrated by the use of geometric figures as shown below. Geometrical proofs are, however, omitted.

(1). $x(a + b + c) = xa + xb + xc$

A rectangle such as xa corresponds to each product.

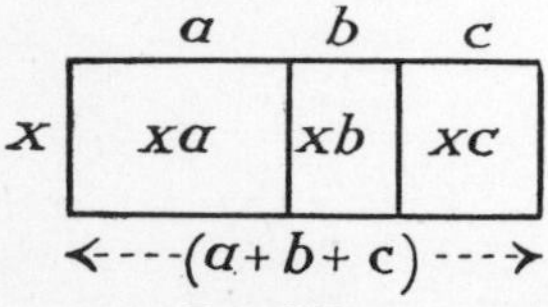

a b c
x | xa | xb | xc
←---(a+b+c)--->

Fig. 263.

The rectangle corresponding to $x(a + b + c)$ is represented by the whole rectangle.

Those corresponding to the products xa, xb, xc, are indicated. The sum of these is obviously equal to the whole rectangle.

The corresponding geometrical theorem is as follows:

If there be two straight lines one of which is undivided and the other is divided into any number of parts, then the rectangle contained by the two straight lines is equal to the sum of the rectangles contained by the undivided line and the parts of the divided line.

(2). $(a + b)^2 = a^2 + b^2 + 2ab$

Area of whole figure $= (a + b)^2$ units of area.

Area of parts $= (a^2 + b^2 + ab + ab)$ units.

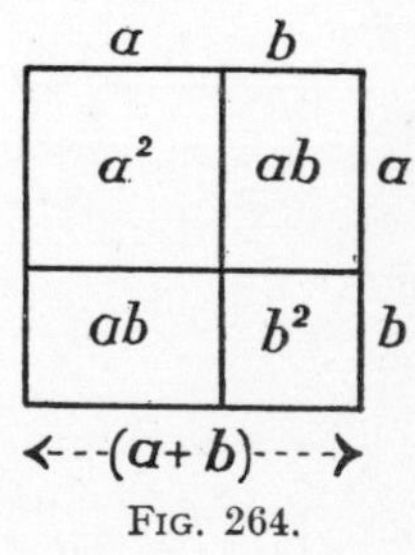

Fig. 264.

Geometrical equivalent. *If a straight line be divided into any two parts, the square on the whole line is equal to the sum of the squares on the two parts plus twice the rectangle contained by these two parts.*

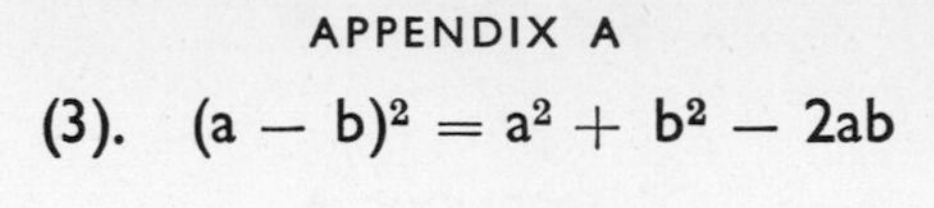

(3). $(a - b)^2 = a^2 + b^2 - 2ab$

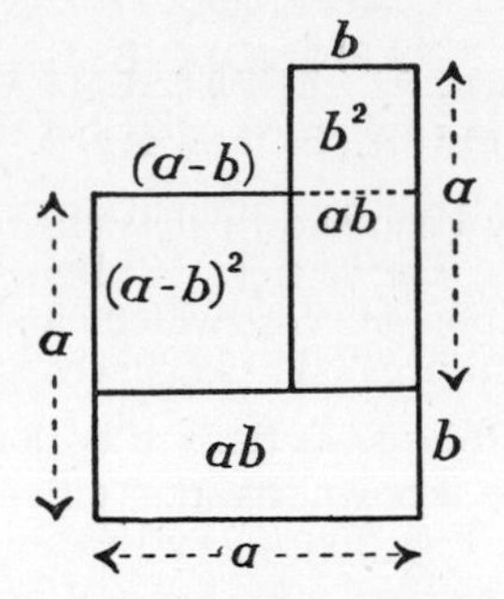

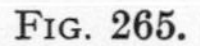

FIG. 265.

Examination of Fig. 265 will make clear that it represents the algebraical identity stated above and the geometrical equivalent which follows.

Geometrical equivalent. *The square on the difference of two straight lines is equal to the sum of the squares on the two lines less twice the rectangle contained by these lines.*

APPENDIX B

SECTIONS OF A CONE AND CYLINDER

It has previously been shown that the normal sections of a cylinder and cone are **circles.** Oblique sections of these solids produce other curves which are of considerable interest and importance. These curves are as follows:

(1) Ellipse. Oblique sections of both **cylinder** and **cone** produce the curve known as the ellipse. Examples of these are shown in Fig. 266 (*a*) and (*b*).

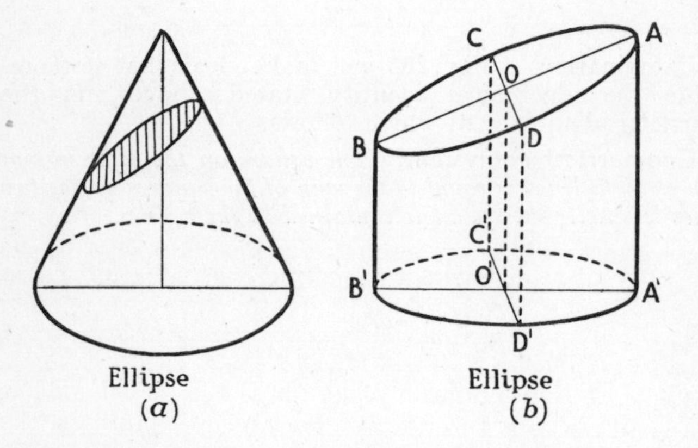

FIG. 266.

The student has probably observed that a perspective view of the circle appears as an ellipse and is drawn as such in the bases of the cone and cylinder not only in Fig. 266, but in all drawings in which they appear in this book.

In the case of the cylinder, whatever the angle made by the plane of the section with the central axis the curve is always a circle or an ellipse. But in the cone the curve may be a circle, ellipse, parabola or hyperbola according to the angle made by the section with the central axis.

When the section meets the curved surface of the solid throughout as in Fig. 266 (*a*) it produces an **ellipse.** But other cases occur as shown below.

(2) The parabola. When the oblique section of the cone is parallel to the slant height, or a generating line, as in Fig. 267 the curve is a **parabola.** *PQ*, which is the axis of the curve (*see* Fig. 112) is parallel to *OA* in Fig. 267 and

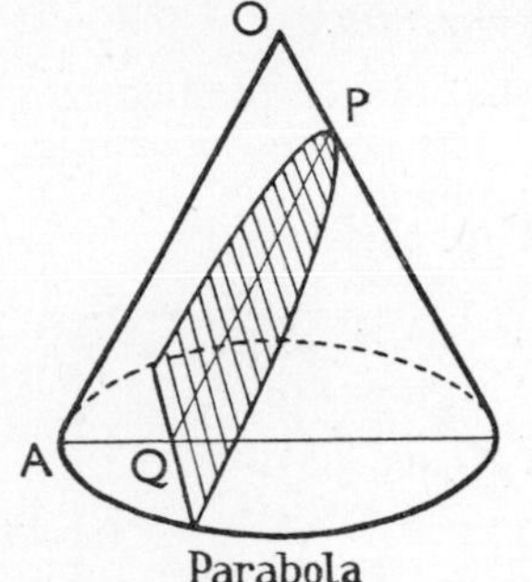

Parabola

FIG. 267.

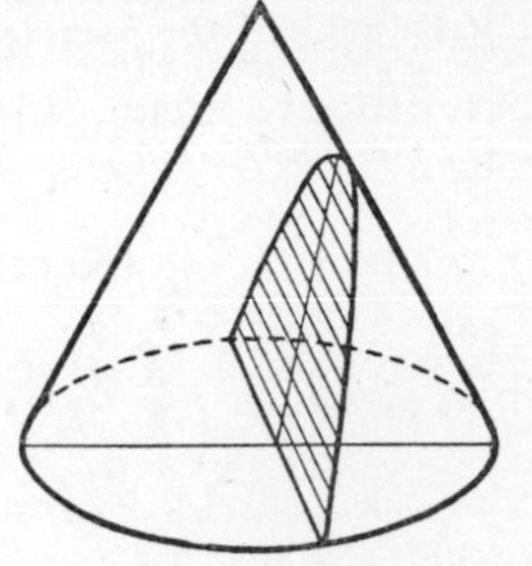

Hyperbola

FIG. 268.

consequently will never meet it. If therefore the cone be regarded as a solid whose generating line is unlimited in length (§ 187) the parabola will stretch to an infinite distance in one direction. The general shape of the curve is indicated in Figs. 110 and 112.

(3) The hyperbola. When the section is oblique but does not conform to either of the above conditions, the curve is a **hyperbola.** It appears as shown in Fig. 268. In the case of the unlimited cone the curve will stretch to an infinite extent and will not be bounded in extent by the curved surface of the solid.

If however the cone is a complete double one (§ 187) the section will evidently cut the other cone.

Consequently there will be two parts or branches of the curve and they will be identical in shape. The form of these is shown in Fig. 113.

ANSWERS

Exercise 1 (p. 42)

1. (1) Acute, (2) obtuse, (3) acute, (4) obtuse, (5) obtuse, (6) adjacent.
2. (*a*) $\frac{4}{5}$ of a right angle, or 72°; (*b*) $\frac{9}{5}$ of a right angle or 162°.
3. (*a*) $52\frac{1}{2}$°, 44° 45′, 17° 20′.
   (*b*) 68°, 25° 30′, 158° 45′.
4. 124°; 56°.
5. BOC; DOB; COA; DOA.
6. Each angle is the complement of $\angle POB$.
7. $A = 40°$; $B = 110°$.
10. (1) 120°; (2) 270°; (3) 720°.
11. (1) 60°; 180°; 25°.

Exercise 2 (p. 54)

1. 45°
2. 90°.
3. W. by N.
4. W.S.W.
5. $112\frac{1}{2}$°.
6. Due E.
7. 32° 40′.
8. 33°.

Exercise 3 (p. 65)

1. 2·3″; 99°; 45° (all approx.).
2. 51°, 59°, 70° (approx.) sum = 180°.
3. 90°; 1·9″, 2·2″.
6. (1) Yes (A); (2) no; (3) no; (4) yes (C); (5) yes (C).
7. $C = 29°$ or 151°.
8. Δs AOD COB are congruent (Theor. A). $\therefore CB = AD$.
9. As in No. 8, $AD = CB$; also Δs AOC, AOD are congruent (Theor. A). $\therefore AC = AD$.
10. Δs AOC, AOD are congruent (Theor. A). $\therefore CO = OD$.

Exercise 4 (p. 75)

1. Corresponding angles: PXB, XYD; PXA, XYC; QYD, YXB; QYC, YXA.
   Alternate angles AXY, XYD; BXY, XYC.
   Interior angles BXY, XYD; AXY, XYC.
2. Following angles are 60°: AXY, XYD, QYC.
   Following angles are 120°: PXA, BXY, XYC, QYD.
3. $PXA = 60°$; $BXY = 60°$; $DYX = 120°$; $QYD = 60°$, $PXB = 120°$.
4. $ABC = 55°$; $BAC = 85°$; $ACB = 40°$.
5. $ACD = 68°$; $ABD = 68°$; $CDB = 112°$; $CAB = 112°$.
6. $\angle ABC$ = corresponding $\angle DQR$ = corresponding $\angle DEF$.
8. 80°.

Exercise 5 (p. 80)

1. (*a*) 58°, (*b*) 53°, (*c*) 44°, (*d*) 112°.
2. 70°, 70°.
3. 72°.
4. 60°.
5. 30°, 60°, 90°.
6. 45°, 45°.
7. 35°, 55°, 35°.
8. 70°, 50°, 60°.
9. 55°, 45°.
11. $A = 74°$, $B = 56°$, $C = 50°$.

Exercise 6 (p. 84)

1. (*a*) $67\frac{1}{2}°$, (*b*) 35°, (*c*) 45°.
2. (*a*) 80°, (*b*) 116°, (*c*) 90°.
3. (*a*) $A = 84°$, $C = 48°$.
   (*b*) B and C each equals 50°.
   (*c*) $B = 70°$, $A = 40°$.
4. 36°, 72°, 72°.
5. 40°, 40°, 100°.
6. 50°, 50°, 80°.
9. 115°.
10. One fourth. All the triangles are congruent.

Exercise 8 (p. 106)

1. (*a*) 70°, 70°, 110°, 110°. (*b*) 72°, 72°, 108°, 108°.
3. 1·4 in.
5. Rectangle.

Exercise 9 (p. 114)

2. 12·5 sq. in.
3. 26·1 sq. cm.
4. (*a*) $54\frac{1}{4}$ sq. ft.; (*b*) 32·495 sq. in.; (*c*) 87·78 sq. cm.
5. 2·48 sq. in.
6. 21·2 sq. cm.
7. 2·52 sq. in.
8. 4·8 cm.
9. 144 sq. ft.
10. 144 sq. ft.

Exercise 10 (p. 120)

1. (*a*) Yes; (*b*) no; (*c*) yes; (*d*) no.
2. (*a*) $\sqrt{2}$ in.; (*b*) $12\sqrt{2}$ in.
3. (*a*) $\frac{\sqrt{3}}{2}$ in.; (*b*) $6\sqrt{3}$ in.
4. 3·47 in.
5. 23·43 miles.
6. 77·5 yds.
7. 42·7 ft.
8. 6 cm.
9. 9·68 in.
10. (*a*) 10·82 sq. in.; (*b*) 10 in.
11. 37·5 ft.
14. 2·6 ft.

Exercise 11 (p. 127)

1. (*a*) $128\frac{4}{7}°$; (*b*) 135°; (*c*) 144°.
2. 18.
3. (*a*) 1 in.; (*b*) $\sqrt{3}$ in.; (*c*) $\frac{3\sqrt{3}}{2}$ sq. in.
5. 9; 140°.
6. 4.

Exercise 12 (p. 141)

1. (*a*) A straight line parallel to AB.
   (*b*) The circumference of a circle of radius 3 ft.
   (*c*) The circumference of a circle, concentric with that of the track.
2. A straight line parallel to the base.
3. A straight line perpendicular to AB and bisecting it.
4. A semi-circle.
5. A straight line parallel to AB.
6. A straight line perpendicular to AB and drawn from B.
7. A straight line, the hypotenuse of a right-angled triangle.
8. The arc of a circle radius BC and centre C.

Exercise 13 (p. 149)

1. (*a*) 31·46 in., (*b*) 7·865 in., (*c*) 5·24 in., (*d*) 3·93 in.
2. 4·188 in.
3. 7·069 ft.
4. (1) 3·1416 in.; (2) 0·785 sq. in.
5. (1) 2·618 in.; (2) 9·59 in.; (3) 10·9 in.
6. 15·5 in.
7. 3·93 ft.
8. 10·6 in.
9. 11·3 ft.
10. 17·9 sq. in.
11. (*a*) 5·71 sq. in.; (*b*) 1·71 sq. in.; (*c*) 98·9689 cm.; (*d*) 10·14 sq. in.
12. 380 acres.

Exercise 14 (p. 155)

1. 1·5 in.
2. 10 in.
3. 24 cm.
4. 3·5 in.
7. A circle concentric with the given circle and of radius 4 in.

Exercise 15 (p. 160)

2. 1·15 in.
3. 45°, 60°, 75°.
9. 95°.

Exercise 16 (p. 168)

1. Two.
3. 2 in.
4. $2\sqrt{7}$ in.
5. A concentric circle of radius $\sqrt{5}$ in.
6. 80°.
9. A concentric circle.
10. A straight line perpendicular to the given straight line at the given point.
11. Two straight lines at right angles to one another and bisecting the given angles.

Exercise 17 (p. 178)

1. $AP = 2$ in., $PB = 3$ in., $AQ = 3\frac{1}{3}$ in., $QC = 4\frac{4}{5}$ in., $BC = 15$ in.
7. 40 sq. in.
8. $\sqrt{2} : 1$.
9. 1 : 2.
10. 1·19.
11.

	Sine.	Cosine.	Tangent
30°	$\frac{1}{2}$	$\frac{\sqrt{3}}{2}$	$\frac{1}{\sqrt{3}}$
45°	$\frac{1}{\sqrt{2}}$	$\frac{1}{\sqrt{2}}$	1
60°	$\frac{\sqrt{3}}{2}$	$\frac{1}{2}$	$\sqrt{3}$

Exercise 18 (p. 184)

1. (*a*) $a^2 = b^2 + c^2 - 2bc \cos A$.
   (*b*) $b^2 = a^2 + c^2 - 2ac \cos B$.
2. (*a*) Area $= \frac{1}{2}bc \sin A$.
   (*b*) Area $= \frac{1}{2}ac \sin B$.
3. $c = 12$.
4. $c = 4$.
5. $\cos C = 0{\cdot}0069$.
6. 19.
7. 19 sq. in. (approx.).
8. 72·4 sq. in. (approx.).

Exercise 19 (p. 189)

1. (*a*) Yes, with 4 axes of symmetry, diagonals and straight line joining bisecting pairs of opposite sides at right angles.
   (*b*) Yes, two axes, the straight lines bisecting pairs of opposite sides at right angles.
   (*c*) Yes; straight line joining the centres.
   (*d*) Yes; radius bisecting the angle of the sector.
   (*e*) No; unless the two non-parallel sides are equal.
   (*f*) Yes; perpendicular bisector of the right angle. Triangle is isosceles.
   (*g*) No.
   (*h*) Yes; straight lines drawn from an angular point perpendicular to the opposite side.
3. Three axes; the straight lines bisecting the angles.
4. Yes; the two diagonals.

Exercise 20 (p. 199)

1. 5 cu. ft.
2. 2897 oz. approx.
3. 110 cu. in.
4. 240 sq. ft.
5. 360 sq. in. approx.
6. 24·73 cu. in.; 1 lb. approx.
7. 18·8 galls. approx.
8. 6·1 sq. in.
9. £4 14*s*. 3*d*.
10. 4·79 in. approx.
11. 49 cu. in. approx.

Exercise 21 (p. 210)

1. (*a*) 144 cu. in.; (*b*) 184·4 sq. in.
2. 15·6 cu. in.
3. (*a*) 205·95 sq. in.; (*b*) 190·9 cu. in.
4. 2·5 sq. in.
5. (1) 17·4 cu. in.; (2) 32·4 c.c.
6. 13·8 ft.
7. (1) 690 ft., (2) 580 ft., (3) 83,500,000 cu. ft. (approx.).
8. 4·39 in.; 36·7 sq. in.
9. 98 sq. ft.
10. (1) 33·5 sq. in., (2) 34·2 cu. in.
11. 38 sq. in.; 28 cu. in.

Exercise 22 (p. 218)

1. 19·7 sq. in.; 8·16 cu. in.
2. £13 17*s*. (approx.).
3. 200,000,000 sq. miles.
4. 4 : 7.
5. 1252 sq. in.
6. 5·9 lb.
7. A double cone, with a common base, and different heights.
8. $\frac{27\pi}{4}$ cu. in.
9. 476 cu. in. (approx.).

Construction No. 16. To construct a triangle equal in area to a given quadrilateral.

Given. $ABCD$ is a quadrilateral.

It is required to construct a triangle equal in area to it.

Construction. Join AC.

From D draw DP parallel to AC and meeting BA produced in P.

Join PC.

PCB is the required triangle.

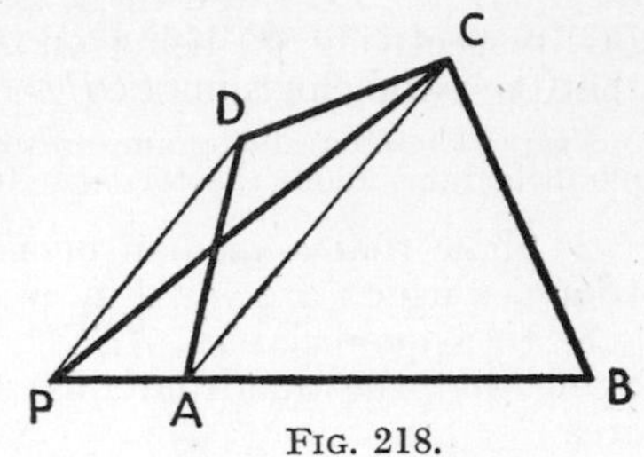

Fig. 218.

Proof. $\triangle ACD = PAC$ in area (*Th.* 30, *Cor.* 2)

To each add the $\triangle ACB$.

$$\therefore \quad \triangle PCB = \text{quad. } ABCD.$$

Corollary. The effect of the above construction is to reduce the number of sides of the given figure by one. This process can be extended to rectilineal figures of any number of sides.

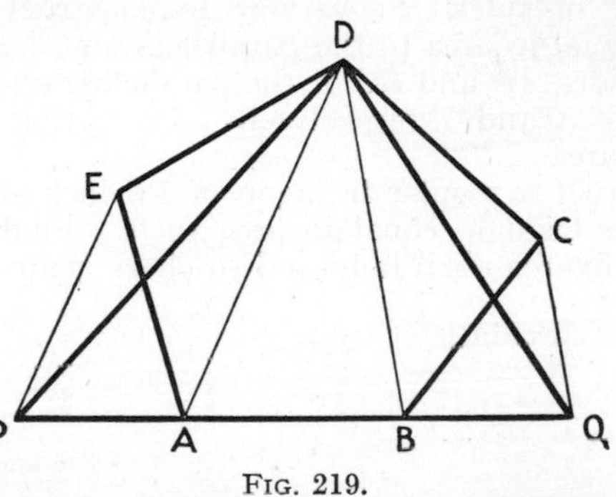

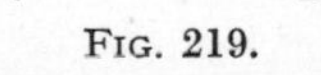

Fig. 219.

For example the pentagon $ABCDE$ (Fig. 219), by the same method as that given above can be reduced to an equivalent quad. $PBCD$.

Then repeating the process, at the next stage this quadrilateral is reduced to the equivalent triangle, PDQ.

Exercise 30

1. $ABCD$ is a parallelogram. Through a point O on the diagonal AC, EOG and FOH are drawn parallel to AB and BC, respectively, E being on AD and F on AB. Prove that the parallelograms $EOHD$, $FOGB$ are equal in area.

Note.—These parallelograms are called the "**complements**" of the parallelograms about the diagonal AC.

2. Show that a median of a triangle divides it into two triangles which are equal in area.

3. In a quadrilateral $ABCD$, O is the mid point of AC. Show that the quadrilaterals $ABOD$, $CBOD$ are equal in area.

4. O is any point on the diagonal BD of the parallelogram $ABCD$. Prove that $\triangle OAB = \triangle OBC$.

5. Prove that the diagonals of a parallelogram divide it into four triangles of equal area.

6. $ABCD$ is a parallelogram. A straight line is drawn parallel to the diagonal BD, and cutting BC and CD in P and Q. Prove that the $\triangle$s ABP, ADQ are equal in area.

7. $ABCD$ is a parallelogram in which AB and CD are the longer pair of sides. Show how to construct on AB a rhombus equal in area to the parallelogram.

8. The sides AB and BC of the parallelogram $ABCD$ are produced to X and Y respectively. Prove that $\triangle CDX = \triangle ADY$ in area.

9. Construct a regular hexagon of 1·5 inch side. Then construct a triangle equal in area to it. Find the areas of the two figures separately and so check your working.